Python for Data Mining Quick Syntax Reference

Valentina Porcu

Apress®

Python for Data Mining Quick Syntax Reference

Valentina Porcu
Nuoro, Italy

ISBN-13 (pbk): 978-1-4842-4112-7   ISBN-13 (electronic): 978-1-4842-4113-4
https://doi.org/10.1007/978-1-4842-4113-4

Library of Congress Control Number: 2018966554

Managing Director, Apress Media LLC: Welmoed Spahr
Acquisitions Editor: Todd Green
Development Editor: James Markham
Coordinating Editor: Jill Balzano

Cover image designed by Freepik (www.freepik.com)

Distributed to the book trade worldwide by Springer Science+Business Media New York, 233 Spring Street, 6th Floor, New York, NY 10013. Phone 1-800-SPRINGER, fax (201) 348-4505, e-mail orders-ny@springer-sbm.com, or visit www.springeronline.com. Apress Media, LLC is a California LLC and the sole member (owner) is Springer Science + Business Media Finance Inc (SSBM Finance Inc). SSBM Finance Inc is a **Delaware** corporation.

For information on translations, please e-mail rights@apress.com, or visit http://www.apress.com/rights-permissions.

Apress titles may be purchased in bulk for academic, corporate, or promotional use. eBook versions and licenses are also available for most titles. For more information, reference our Print and eBook Bulk Sales web page at http://www.apress.com/bulk-sales.

Any source code or other supplementary material referenced by the author in this book is available to readers on GitHub via the book's product page, located at www.apress.com/9781484241127. For more detailed information, please visit http://www.apress.com/source-code.

Printed on acid-free paper

Table of Contents

About the Author

Valentina Porcu is a computer geek with a passion for data mining and research, and a PhD in communication and complex systems. She has years of experience in teaching in universities in Italy, France, and Morocco—and online, of course! She works as a consultant in the field of data mining and machine learning, and enjoys writing about new technologies and data mining. She spent the past nine years working as freelancer and researcher in the field of social media analysis, benchmark analysis, and web scraping for database building, in particular in the field of buzz analysis and sentiment analysis for universities, startups, and web agencies across the United Kingdom, France, the United States, and Italy. Valentina is the founder of Datawiring, a popular Italian data science resource.

About the Technical Reviewer

Karpur Shukla is a research fellow at the Centre for Mathematical Modeling at Flame University in Pune, India. His current research interests focus on nonequilibrium fluctuation theorems for models of topological quantum field theories (with application to topological quantum computing) and models of reversible computing. He received an MS in physics from Carnegie Mellon University, with a background in theoretical analysis of materials for spintronics applications as well as Monte Carlo simulations for the renormalization group of finite-temperature spin lattice systems.

Introduction

Translated by Nicola Menicacci

Python is an interpreted, interactive, and object-oriented language. It features a library of functions, is extendable (as it can be used to create new modules easily), and is available for all operating systems. For these and other reasons, it is also one of the most used programming languages when it comes to data mining and machine learning.

My goal is to accompany you as you start to study this programming language, show you basic concepts, and then help you move on to data mining. We'll begin by looking at how to use Python and its structures, how to install Python, and how to determine which tools are best suited for data analysis, and then switch to an introduction to data mining packages. *Python for Data Mining Quick Syntax Reference* is an introductory book. It provides guidance—from taking your first programming steps with Python, to manipulating and importing datasets, to examining examples of data analysis. It does not explain fully topics such as machine learning and statistics using Python, which are beyond the scope of this volume.

Who This Book Is For

This book is intended for those of you who want to gain a better understanding of the Python programming language from a data analysis perspective. We will start by reviewing Python's basic concepts, then focus on the most used packages for data analysis. To download the code, to delve more deeply into some topics, and to acquire more practical information about Python and data mining, please visit my website (**Datawiring.me**). From the site's home page, you can subscribe to my newsletter to receive updates about the latest in Python coding and other news. My advice for those of you who are beginning programmers is to write the code manually to gain a greater understanding of it.

How This Book Is Organized

Python for Data Mining Quick Syntax Reference consists of 11 chapters. In Chapter 1, we look at some basic installation concepts and the tools available for programming in Python. We also examine differences between Python2 and Python3 and learn how to set up a work folder.

In Chapter 2, we study some basic concepts about creating objects, entering comments, and reserving words for the system; and look at the various types of operators that are part of the grammar of the Python programming language.

In Chapter 3, we extend our work with basic Python structures—such as tuples, lists, dictionaries, sets, strings, and files—and learn how to create and convert them.

In Chapter 4, we create small, basic functions and learn how to save them.

Chapter 5 deals with conditional instructions that allow us to extend the power of a function. In addition, we review other important functions as well.

In Chapter 6, we investigate basic concepts related to object-oriented programming and examine the concepts of modules, methods, and error handling.

Chapter 7 is dedicated to importing files using some of the basic features we have learned. We learn how to open and edit text files in .csv format, in addition to various other formats.

Chapters 8 through 11 explain Python's most important data mining packages: NumPy and SciPy for mathematical functions and random data generation, pandas for dataframe management and data import, Matplotlib for drawing charts, and scikit-learn for machine learning. With regard to scikit-learn, the discussion is limited to basic coverage of the code of the various algorithms. Because of the complexity of the topic, we do not examine the details for the various techniques.

CHAPTER 1

Getting Started

Python is one of the most important programming languages used in data science. In this chapter, you'll learn how to install Python and review some of the integrated development environments (IDEs) used for data analysis. You'll also learn how to set up a working directory on your computer.

Installing Python

Python2 and Python3 can be downloaded easily from `https://www.python.org/downloads/` (Figure 1-1) and then installed. Note that if you are working on a Unix system using a Mac or Linux, Python is preinstalled. Simply type "python" to load the program.

Figure 1-1. *Python home page*

© Valentina Porcu 2018
V. Porcu, *Python for Data Mining Quick Syntax Reference*,
https://doi.org/10.1007/978-1-4842-4113-4_1

From the python.org (http://python.org/) website, click Downloads then select the appropriate version to use based on your operating system. Then, follow the on-screen instructions to install Python.

Editor and IDEs

There are many ways to use a programming language such as Python. To start, type the word "python" followed immediately by its version number. There is no space before the number. For example, in Figure 1-2, I've typed "python2."

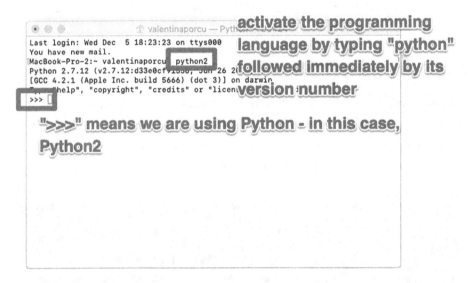

Figure 1-2. *Terminal with Python open*

Writing code this way may prove to be somewhat cumbersome, so we use text editors or IDEs to facilitate the process.

There are many editors (those that are free and those that can be purchased) that differ in their completeness, scalability, and ease of use. Some are simple and some are more advanced. The most used editors include Sublime Text, Text Wrangler (`http://www.barebones.com/`), Notepad++ (`http://notepad-plus-plus.org/download/v7.3.1.html`) (for Windows), or TextMate (`http://macromates.com/`) (for Mac).

As for Python-specific IDEs, Wingware (`http://wingware.com/`), Komodo (`http://www.activestate.com/komodo-ide`), Pycharm, and Emacs (`http://www.gnu.org/software/emacs/`) are popular, but there are plenty of others. They provide tools to simplify work, such as self-completion, auto-editing and auto-indentation, integrated documentation, syntax highlighting, and code folding (the ability to hide some pieces of code while you works on others), and to support debugging.

Spyder (which is included in Anaconda (`http://www.continuum.io/downloads`)) and Jupyter (`http://jupyter.readthedocs.io/en/latest/`), that you can download from the website `www.anaconda.com`, are the IDEs used most in data science, along with Canopy. A useful tool in Jupyter is nbviewer, which allows the exchange of Jupyter's .ipynb files, and can be downloaded from `http://nbviewer.jupyter.org`. nbviewer can also be linked to GitHub.

As for Anaconda, which is a very useful tool because it also features Jupyter, it can be downloaded from `http://www.continuum/`. A partial list of resources installed with Anaconda (which contains more than 100 packets for data mining, math, data analysis, and algebra) is presented in Figure 1-3. You can view the complete list by opening the a terminal window shown in Figure 1-3 and then typing:

```
conda list
```

```
# packages in environment at /Users/valentinaporcu/anaconda:
#
_license                1.1                      py35_1
_nb_ext_conf            0.3.0                    py35_0
alabaster               0.7.9                    py35_0
alembic                 0.8.8                     <pip>
anaconda                custom                   py35_0
anaconda-clean          1.0.0                    py35_0
anaconda-client         1.5.1                    py35_0
anaconda-navigator      1.3.1                    py35_0
appnope                 0.1.0                    py35_0
appscript               1.0.1                    py35_0
argcomplete             1.0.0                    py35_1
astroid                 1.4.7                    py35_0
astropy                 1.2.1                 np111py35_0
babel                   2.3.4                    py35_0
backports               1.0                      py35_0
beautifulsoup4          4.5.1                    py35_0
bitarray                0.8.1                    py35_0
blaze                   0.10.1                   py35_0
bokeh                   0.12.2                   py35_0
boto                    2.42.0                   py35_0
bottleneck              1.1.0                 np111py35_0
cffi                    1.7.0                    py35_0
chardet                 2.3.0                    py35_0
chest                   0.2.3                    py35_0
click                   6.6                      py35_0
cloudpickle             0.2.1                    py35_0
clyent                  1.2.2                    py35_0
colorama                0.3.7                    py35_0
conda                   4.3.11                   py35_0
conda-build             2.0.2                    py35_0
conda-env               2.6.0                        0
configobj               5.0.6                    py35_0
```

Figure 1-3. *Part of the resources installed with Anaconda*

We can program with Python using one or more of these tools, depending on our habits and what we want to do. Spyder (Figure 1-4) and Jupyter (Figure 1-5) are very common for data mining. Both can be used and installed individually. For example, Jupyter can be tested using http://try.jupyter.org/. However, both Spyder and Jupyter are available after Anaconda is installed.

Figure 1-4. *Spyder home screen*

Figure 1-5. *Example of open script on Jupyter IDE*

Python code can be run directly from a computer terminal or saved as a .py file and then run from these other editors. As mentioned earlier, ">>>" (displayed in Figure 1-6) tells us we are running Python code.

```
Python 2.7.12 (v2.7.12:d33e0cf91556, Jun 26 2016, 12:10:39)
[GCC 4.2.1 (Apple Inc. build 5666) (dot 3)] on darwin
Type "help", "copyright", "credits" or "license" for more information.
>>>
```

Figure 1-6. *The command prompt in Python*

To follow the examples presented in this book, I recommend you install Anaconda (Figure 1-7) from the **AAnaconda.com** web site and use Jupyter. Because Anaconda automatically includes (and installs) a set of packages and modules that we will use later, we won't have to install packages or modules separately thereafter; we'll already have them loaded and ready to use.

Figure 1-7. *Anaconda's main screen*

Differences between Python2 and Python3

Python was released in two different versions: Python2 and Python3. Python2 was born in 2000 (currently, the latest release is 2.7) and its support is expected to continue until 2020. It is the historical and most complete version.

Python3 was released in 2008 (current version is 3.6). There are many libraries in Python3, but not all of them have been converted from Python2 for Python3.

The two versions are very similar but feature some differences. One example includes mathematical operations:

Listing 1-1. Mathematical Operations in Python 2.7

```
>>> 5/2
2

# Python2 performs division by breaking the decimal.
```

Listing 1-2. Mathematical Operations in Python 3.5.2

```
>>> 5/2
2.5
```

To get the correct result in Python2, we have to specify the decimal as

```
>>> 5.0/2
2.5

# or like this

>>> 5/2.0
2.5

# or specify we are talking about a decimal (float)

>>> float(5)/2
2.5
```

7

To keep the two versions of Python together, you can also import Python into a form called *future*, which allows you to import Python3 functions into Python2:

```
>>> from __future__ import division

>>> 5/2

2.5
```

For a closer look at the differences between the two versions of Python, access <u>this online resource</u> (http://sebastianraschka.com/ Articles/2014_python_2_3_key_diff.html).

Why choose one version of Python over the other? Python2 is the best-defined and most stable version, whereas Python3 represents the future of the language, although the two versions may not always coincide. In the first part of this book, I highlight the differences between the two versions. However, beginning with Chapter 7 and moving to the end of the book, we will use Python3.

Let's start by setting up a work directory. This directory will house our files.

Work Directory

A work directory stores our scripts and our files. It is where Python automatically looks when we ask it to import a file or run a script. To set up a work directory, type the following in the Python shell:

```
>>> import os

>>>> os.getcwd()
'~/mypc'

# to edit the work directory, we use the following notation,
inserting the new directory in parentheses
```

```
>>> os.chdir("/~/Python_script")
```

```
# then we determine whether it is correct
```

```
>>> os.getcwd()
'~/Python_script'
```

Now, when we want to import a file in our workbook, we simply type the name of the file followed by the extension, all surrounded by double quotation marks:

```
"file_name.extension"
```

For instance,

```
"dataframe_data_collection1.csv"
```

Python checks whether there is a file with that name inside that folder and imports it. The same thing happens when we save a Python file by typing it on a computer. Python automatically puts it in that folder. Even when we run a Python script, as we will see, we have to access the folder where the script (the work directory or another one) is located directly from the terminal.

If we want to import a file that is not in the work directory but is elsewhere on our computer or on the Web, we do this by entering the full file address:

```
"complete_address.file_name.extension"
```

For instance,

```
"/~/dataframe_data1.csv"
```

Now let's make sure that you understand the difference between using a the terminal and starting a session in our favorite programming language.

Using a Terminal

To run Python scripts, we first open a terminal window, as shown in Figure 1-8.

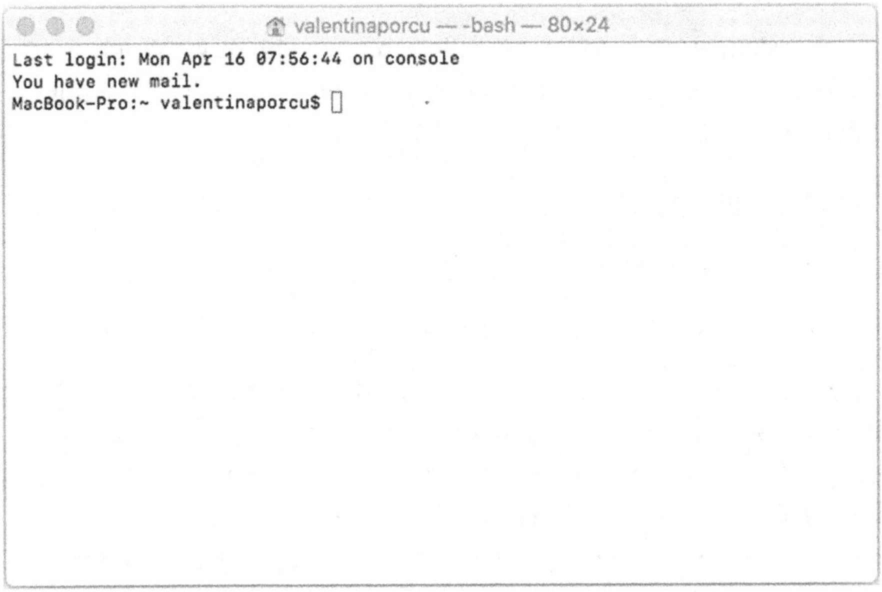

Figure 1-8. *My terminal*

As you can see, the dollar symbol ($) is displayed, not the Python shell symbol (>>>). To view a list of our folders and files, use the "ls" command (Figure 1-9).

Figure 1-9. *List of resources on my computer*

At this point, we can move to the Python_test folder by typing

```
cd Python_test
```

In that folder, I find my Python scripts—that is, the .py files I can run by typing

```
python test.py
```

test.py is the name of the script I am going to run.

Summary

In this chapter we learned how to install Python and I reviewed some of the various IDEs we can use for data analysis. We also examined Python2 and Python3, and learned how to set up a work directory on a terminal.

CHAPTER 2

Introductory Notes

In this chapter we examine Python objects and operators, and learn how to write comments in our code. Including comments in your code is very important for two reasons. First, they serve as a reminder of our thought processes on the work we did weeks or months after we've created a script. Second, they help other programmers understand why we did what we did.

Objects in Python

In Python, any item is considered an object, a container to place data. Python objects include tuples, lists, sets, dictionaries, and containers. Python processing is based on objects.

Each object is distinguished by three properties:

1. A name

2. A type

3. An ID

Object names consist of alphanumeric characters and underscores—in other words, all characters from A through Z, a through z, 0 through 9, and _. "Type" is the type of object, such as string, numeric, or Boolean. "ID" is a number that identifies the object uniquely.

© Valentina Porcu 2018
V. Porcu, *Python for Data Mining Quick Syntax Reference*,
https://doi.org/10.1007/978-1-4842-4113-4_2

Objects remain inside computer memory and can be retrieved. When no longer needed, a garbage-collecting feature frees up the space.

Some IDs and types are assigned automatically. With regard to names, however, note there are some names that cannot be used for objects.

Reserved Terms for the System

Python has a set of words that is reserved for the system and cannot be used in names for objects or functions. These words are the following:

and, as, assert, break, class, continue, def, del, elif, else, except, exec, False, finally, for, from, global, if, import, in, is, lambda, None, not, or, pass, print, raise, return, True, try, while, with, yield

In addition, object names in R are subject to some rules:

- They must begin with a letter or underscore.

- They must contain only letters, numbers, and underscores.

- They are case sensitive. For example, a test object is not the same as a TEST object or a Test object.

Entering Comments in the Code

In Python, any comment preceded by the # symbol is not read by the program as code; it is ignored. This feature is useful for commenting on code and refreshing our memories later. Comments can be written both in the code and to the side.

```
# comment no. 1
>>> print("Hello World") # comment no. 2
```

To write a comment on multiple lines, we can also use three quotation marks, like this:

```
"""
    comment line 1
    comment line 2
    comment line 3
"""
```

Types of Data

Python data are various types and are summarized in Table 2-1.

Table 2-1. *Python Data Types*

Data Type	Example
Integer (int)	1, 20, −19
Float	1.7, 12.54
Complex	657.23e+34
String (str)	"Hello World", 'string1',""" string2 """
List	list = ['a', 'b', 'c']
Tuple	tuple = ('Laura', 29, 'Andrea', 4)
Dictionary	dictionary = {'name' : 'Simon', 'key': 'D007'}

To determine object type, we use the type() function:

```
# we create an x object

>>> x = 1
>>> type(x)
<class 'int'>

# a y object

>>> y = 20.75
>>> type(y)
<class 'float'>

# and a z object

>>> z = "test"
>>> type(z)
<class 'str'>
```

File Format

After you create a script in Python, you need to save it with a .py extension. Typically, when it comes to complex scripts, you create a script on an editor that you then run. A .py script can be written using any one of the different editors discussed earlier—even a normal text editor—and then can be renamed with .py extension.

Operators

Python includes a series of operators that are divided into several groups:

- Mathematical

- Comparison

- Membership

- Bitwise

- Assignment

- Logical

- Identity

Beside these operators, there is also a hierarchy that marks the order in which they are used. Operators are very important for arithmetic, or even to extract some data with specific characteristics.

Mathematical Operators

When we open Python, the simplest thing we can do is use it to perform mathematical operations, for which we use the mathematical operators shown in Table 2-2.

Table 2-2. *Mathematical Operators*

Operator	Description	Example
+	Addition	3 + 2 = 5
−	Subtraction	10 − 4 = 6
*	Multiplication	4 * 3 = 12
/	Division	20/2 = 10
%	Modulo	21/2 = 1
**	Exponentiation	3 ** 2 = 9
//	Floor	10.5 // 2 = 5.0

Let's open Python and perform some mathematical operations:

```
>>> 10+7
17

>>> 15-2
13

>>> 2*3
6

>>> 10/2
5

>>> 3**3
27

>>> 10/3
3

>>> 25//7
3
```

Comparison and Membership Operators

In Python we also have comparison operators (comparators) and membership operators (Table 2-3).

Table 2-3. *Comparison and Membership Operators*

Operator	Description
>	Greater than
<	Less than
==	Equal to
>=	Greater than or equal to
<=	Less than or equal to
!=	Different
is	Identity
is not	Non identity
in	Exists in
not in	Does not exist in

These operators are used to test relationships between objects. For example,

```
# we create two objects

>>> x = 5
>>> y = 10

# let us verify that x is greater than y

>>> x > y
False

# the output is a logical vector that tells us x is not greater
than y
```

```
# let us see if x is less than y

>>> x < y
True
# this time the answer is affirmative

# we create another object, z, with the same value as x

>>> z = 5
# let us verify with the equality operator whether z is equal to x

>>> z == x
True
# in this case, the output is positive

# let us verify whether z is different from y

>>> z != y
True
# in this case, the output is also positive

# let us create a tuple

>>> v1 = (1,2,3,4,5,6,7)
# and verify whether the number 2 is in the tuple

>>> 2 in v1
True

# let us verify whether the number 8 is NOT in tuple v1

>>> 8 not in v1
True

# let us verify whether the number 7 is NOT in tuple v1

>>> 7 not in v1
False
```

Python compares text strings lexicographically using, for example, the ASCII value of the characters. We cannot compare strings and numbers because we would throw an error.

Bitwise Operators

The bitwise operators shown in Table 2-4 are useful when specifying more than one condition. An example is when we need to extract data from an object, such as a dataset.

Table 2-4. *Bitwise Operators*

Operator	Description
& or **and**	And
| or **or**	Or
^	Xor
~	Bitwise not
<<	Left shift
>>	Right shift

Bitwise operators can be used together with comparators. Let's look at some examples:

```
>>> 3 < 4 and 4 > 3
True

# and also

>>> 3 < 4 & 4 > 3
True
```

```
# here is an example with or (|)

>>> 3 < 4 | 4 > 3
True

# at least one of the statements must be valid

>>> 3 == 4 or 4 > 3
True
```

Assignment Operators

Assignment operators are used to assign a name to a given object (Table 2-5).

Table 2-5. *Assignment Operators*

Operator	Description	Example
=	Basic assignment operator	x = 5 + 6
+=	Adds an element and assigns the result to the name	x += y (corresponds to x = x + y)
-=	Subtracts an element and assigns the result to the name	x -= y (corresponds to x = x − y)
/=	Divides an element and assigns the result to the name	x /= y (corresponds to x = x/y)
*=	Multiplies an element and assigns the result to the name	x *= y (corresponds to x = x * y)
%=	Modulo and reassignment	x %= y (corresponds to x = x % y)
**=	Exponentiation and reassignment	x **= y (corresponds to x = x ** y)
//=	Floor division and reallocation	x//=y (corresponds to x = x//y)

Let's look at some examples:

```
# we create an x object with the value 10

>>> x = 10

# sum x and overwrite the variable x again with the same name

>>> x = x + 5

>>> x

15

# let's try "+="

>>> x += 5

>>> x

20

# and now "-="

>>> x -= 5

>>> x

15

# now let's use the operator "*="

>>> x *= 3

>>> x

45
```

```
# and the operator "/="

>>> x /= 3

>>> x

15

# then the operator "**="

>>> x **= 2

>>> x

225

# and finally the operator "//="

>>> x //= 2

>>> x

112
```

Each time, Python performs the operation and records the result in the x object.

Operator Order

When it comes to mathematical operators, we must be aware that there is an order priority that must be observed when case brackets are not inserted. This is similar to mathematical operations in which multiplication takes precedence over addition. Table 2-6 lists some of the priority rules that govern the order of operations.

Table 2-6. *Priority Rules for Operators*

Operator	Priority (highest to lowest)
**	Exponentiation has the highest priority
-	Denial
* / // %	Multiplication, division, modulo, floor division
+ -	Addition and subtraction
>> <<	Bitwise right and left
&	Bitwise AND
^ |	Bitwise OR
<=, >, <, >=	Less than, more than, smaller, bigger than
== !=	Equal, different
= += -= *= /= %= **= //=	Assignment operators
is / is not	Comparison
in / not in	Comparison
not / or / and	Comparison

Indentation

Python uses indentation to limit blocks of instructions, which makes the code more readable. Code blocks are thus defined by indentation. Typically an indentation corresponds to four spaces.

Let's look at an example of indentation in a function:

```
>>> def multiply_xy(x, y):
...     "'let's multiply x and y
...     "'
...     return(x*y)
>>> multiply_xy(5,6)
30
```

Quotation Marks

Quotation marks in R are used primarily to define strings. They can be single, double, or triple. Triple quotation marks are used to wrap words and insert comments on multiple lines. An example of this is when we wish to include documentation within a function we are creating.

```
>>> ex1 = 'single quote'

>>> ex2 = "double quote"

>>> ex3 = """
        text string 1
        text string 2
        text string 3
"""
```

We examine string management in Chapter 5.

Summary

In this chapter, we studied commenting and operators. When we write code for data analysis, it is important that we include comments not only as a reminder to ourselves but also as a guide for other programmers. Operators, and their ranking, help us create and define our code.

CHAPTER 3

Basic Objects and Structures

One of the most important features of Python is managing data structures. Let's take a look at them.

Numbers

The numbers in Python can be any of the following:

- Integers, or int

- Floating points, or float

- Complex

- Booleans—that is, True or False

Let's look at some examples:

```
# create an object containing an integer (int)
>>> n1 = 19
>>> type(n1)
<type 'int'>
```

© Valentina Porcu 2018
V. Porcu, *Python for Data Mining Quick Syntax Reference*,
https://doi.org/10.1007/978-1-4842-4113-4_3

```
# a float
>>> n2 = 7.5
>>> type(n2)
<type 'float'>
# a Boolean (True/False)
>>> n3 = True
>>> type(n3)
<type 'bool'>
# a complex number
>>> n4 = 3j
>>> type(n4)
<type 'complex'>
```

Container Objects

At the heart of Python are the various types of objects that can be created (Table 3-1).

Table 3-1. *Python Container Objects*

Container	Delimited by
Tuples	()
Lists	[]
Dictionaries	{}
Sets	{}
Strings	" " " "

Let's examine each of them in turn.

Tuples

The tuples, as well as strings and lists, are part of the *sequence* category. Sequences are iterative objects that represent arbitrary-length containers. Tuples are sequences of heterogeneous and immutable objects, and are identified by parentheses. The fact that they are immutable means that after we have created a tuple, we cannot alter it; we cannot replace one of its elements with another. Tuples are very efficient with regard to memory consumption and runtime.

Let's create a tuple:

```
>>> t1 = (1,2,3,4,5)

# we interrogate with the type() function based on the object
type we created

>>> type(t1)
<class 'tuple'>

# Python tells us we created a tuple, so we have created the
right data structure
```

Common operations for sequences are indexing and slicing, and concatenation and repetition. As mentioned, we cannot modify a tuple after it has been created, but we *can* extract, concatenate, or repeat its elements.

```
# we create another tuple

>>> t2 = ("a", "b", "c", "d")
>>> type(t2)
<class 'tuple'>

# we extract the first element of tuple t2

>>> t2[0]
'a'
```

```
# to count the elements of a tuple, we start with zero; to
extract "a", which is the first element, we use square brackets
for slicing and insert the number 0 between them
```

```
# we can also use the minus sign to extract elements of a
tuple; these elements are counted from the last to the first

>>> t2[-1]
'd'
```

```
# we can extract more than one element using a colon

>>> t2[1:3]
('b', 'c')
```

To determine whether an item is present in a tuple, we use the "in" operator:

```
>>> 'a' in t2
True
```

```
>>> 'z' in t2
False
```

As mentioned, tuples are immutable. If we try to replace one element of a tuple with another, we get an error message:

```
>>> t2['a'] = 15
Traceback (most recent call last):
  File "<stdin>", line 1, in <module>
TypeError: 'tuple' object does not support item assignment
```

To display the functions available for tuples, type

```
>>> dir(t2)
['__add__', '__class__', '__contains__', '__delattr__',
'__dir__', '__doc__', '__eq__', '__format__', '__ge__',
'__getattribute__', '__getitem__', '__getnewargs__', '__gt__',
'__hash__', '__init__', '__iter__', '__le__', '__len__', '__
lt__', '__mul__', '__ne__', '__new__', '__reduce__', '__reduce_
ex__', '__repr__', '__rmul__', '__setattr__', '__sizeof__',
'__str__', '__subclasshook__', 'count', 'index']
```

We can add elements to tuples by using the functions available for them:

```
>>> t2 = t2.__add__(('xyz',))

# let's see our tuple again

>>> t2
('a', 'b', 'c', 'd', 'xyz')
```

Last, we can create tuples that contain more than one type of object:

```
>>> t3 = (1,2,3,4,5, "test", 20.75, "string2")

>>> t3
(1, 2, 3, 4, 5, 'test', 20.75, 'string2')
```

Lists

Python lists include items of various types. They are similar to tuples, with the difference that they are mutable; you can add or delete items from a list.

To create a list we include its elements in square brackets, separated by a comma:

```
>>> list1 = ["jan", "feb", "mar", "apr"]

type(list1)
<class 'list'>
```

We can also create lists that contain numerical, logical, or string values, or we can mix multiple data types:

```
>>> list2 = ["one", 25, True]

>>> type(list2)
<class 'list'>
```

We can display a list using the print() function:

```
>>> print(list1)
['jan', 'feb', 'mar', 'apr']
```

Or we can determine its length with the len() function:

```
>>> len(list1)
4
```

We can also print a single list item according to is location:

```
>>> list1[0]
'jan'

>>> list1[-2]
'mar'
```

```
# if we insert a position that does not match any item in the
list, we get an error
>>> list1[7]
Traceback (most recent call last):
  File "<stdin>", line 1, in <module>
IndexError: list index out of range
```

We can select some items from a list:

```
>>> list1 = ["jan", "feb", "mar", "apr"]

>>> list1[1:]

['feb', 'mar', 'apr']

>>> list1[:3]

['jan', 'feb', 'mar']
```

We can multiply a list:

```
>>> list1*2
['jan', 'feb', 'mar', 'apr', 'jan', 'feb', 'mar', 'apr']
```

Or we can create a new list by combining two lists:

```
>>> list3 = list1 + list2

>>> list3
['jan', 'feb', 'mar', 'apr', 'one', 25, True]
```

We can even extract some items and save them to another list, which really means we are talking about slicing.

```
>>> list4 = list3[2:6]

>>> list4
['mar', 'apr', 'one', 25]
```

We can also delete an item from a list like this:

```
>>> del list1[1]
```

```
>>> list1
['jan', 'mar', 'apr']
```

By typing the dir() function with a list, we can see all the operations we can do on that list:

```
>>> dir(list1)
['__add__', '__class__', '__contains__', '__delattr__', '__
delitem__', '__dir__', '__doc__', '__eq__', '__format__', '__
ge__', '__getattribute__', '__getitem__', '__gt__', '__hash__',
'__iadd__', '__imul__', '__init__', '__iter__', '__le__', '__
len__', '__lt__', '__mul__', '__ne__', '__new__', '__reduce__',
'__reduce_ex__', '__repr__', '__reversed__', '__rmul__',
'__setattr__', '__setitem__', '__sizeof__', '__str__', '__
subclasshook__', 'append', 'clear', 'copy', 'count', 'extend',
'index', 'insert', 'pop', 'remove', 'reverse', 'sort']
```

Some of the most important functions include the following:

- **append()** Adds elements to our list

- **clear()** Removes all items in a list

- **copy()** Makes a copy of the list

- **extend()** Combines two lists

- **insert()** Adds an item to a specific location in the list

- **pop()** Removes an item from the list

- **remove()** Removes an item from a specific location of the list

Let's use some of these operations on list3.

```
>>> list3
['jan', 'feb', 'mar', 'apr', 'one', 25, True]

# we can add an element with the append() method

>>> list3.append(7)
>>> list3
['jan', 'feb', 'mar', 'apr', 'one', 25, True, 7]


# reverse the order of the list items with reverse()

>>> list3.reverse()
>>> list3
[7, True, 25, 'one', 'apr', 'mar', 'feb', 'jan']

# delete the last element with pop()

>>> list3.pop()
'jan'

>>> list3
[7, True, 25, 'one', 'apr', 'mar', 'feb']


# reorder items of a list in ascending order with sort()

>>> list5 = [100, 12, 45, 67, 89, 7, 19]
>>> list5.sort()
>>> list5
[7, 12, 19, 45, 67, 89, 100]

# extend a list with another list with extend()

>>> list5.extend([260, 35, 98, 124])

>>> list5
[7, 12, 19, 45, 67, 89, 100, 260, 35, 98, 124]
```

```
# last, we can delete items in a list with the clear function
>>> list5.clear()
>>> list5
[]
```

We can also create lists that contain sublists:

```
>>> list6 = [(5,7), (9,2), (2,3), (14,27)]
>>> list6
[(5, 7), (9, 2), (2, 3), (14, 27)]
# in this case, let's select the third element of the list6
object:
>>> list6[2]
(2, 3)
# let's select only the second element of the third element of
list6:

>>> list6[2][1]
3
```

We can create a list that features a series of numbers by using the range() function.

```
# the range function()creates a list of numbers from 1 to 19:

>>> list7 = range(20)
# let us check the type of object
>>> type(list7)
<type 'list'>
```

```
# we print the object
>>> print(list7)
[0, 1, 2, 3, 4, 5, 6, 7, 8, 9, 10, 11, 12, 13, 14, 15, 16, 17,
18, 19]
```

Dictionaries

Another Python data structure includes dictionaries. They are containers that store key-value pairs and are distinguished by the use of braces and two points. Dictionaries are mutable but cannot be ordered. We cannot extract items from a dictionary as we did with lists and tuples.

In our first example, let's look at a dictionary that records the names and heights of subjects:

```
>>> dict1 = {'Laura': 163, 'Francis': 169, 'Kate': 165}

>>> type(dict1)
<type 'dict'>
```

We can query the dictionary for a given value:

```
>>> print dict1['Francis']
169
```

We can also add an element to our dictionary and rewrite it:

```
>>> dict1['Simon'] = '180'

>>> dict1
{'Laura': 163, 'Simon': '180', 'Francis': 169, 'Kate': 165}
```

To list dictionary keys, we use the .keys method:

```
>>> dict1.keys()
['Laura', 'Simon', 'Francis', 'Kate']
```

To get only the values, we use the .values method:

```
>>> dict1.values()
[163, 169, 165]
```

To determine whether a given key is in our dictionary, we use the "in" operator:

```
>>> 'Laura' in dict1
True

>>> 'Stephanie' in dict1
False
```

We can delete a dictionary element with the del command:

```
>>> del dict1['Simon']

>>> dict1
{'Laura': 163, 'Francis': 169, 'Kate': 165}
```

We can delete all dictionary elements with the .clear method:

```
>>> dict1.clear()

>>> dict1
{}
```

Now, let's create another dictionary:

```
>>> dict2 = {'Statistics':28, 'Machine Learning':30,
'Marketing':27,'Analysis':29}

>>> dict2
{'Marketing':27, 'Statistics':28, 'Analysis':29,
'Machine Learning':30}
```

We can verify the number of elements that make up the dictionary with len():

```
>>> len(dict2)
4
```

Dictionary dict2 features four key-value pairs.

We can query a dictionary about a given element even without the print() function:

```
>>> dict2['Marketing']
27
```

Let's check the keys with the list() function:

```
>>> list(dict2)
['Marketing', 'Statistics', 'Analysis', 'Machine Learning']
```

We can place the keys in alphabetical order:

```
>>> sorted(list(dict2))
['Analysis', 'Machine Learning', 'Marketing', 'Statistics']
```

We can display values only with the .values method:

```
>>> dict2.values()
[27, 28, 29, 30]
```

And can we display all items with the .items method:

```
>>> dict2.items()

[('Marketing', 27), ('Statistics', 28), ('Analysis', 29),
('Machine Learning', 30)]
```

We can list the elements in our dictionary by creating a function:

```
>>> for i in dict2: print(i)
...
Marketing
Statistics
Analysis
Machine Learning
```

We can also delete one of the items with the .pop method:

```
>>> dict2.pop('Marketing')
27

>>> dict2
{'Statistics': 28, 'Analysis': 29, 'Machine Learning': 30}
```

The .popitem method, on the other hand, deletes a random element from the dictionary:

```
>>> dict2.popitem()
('Statistics', 28)

>>> dict2
{'Analysis': 29, 'Machine Learning': 30}
```

There are now two elements in the dictionary dict2. We can update one of the values—for example, 29—by subtracting:

```
>>> dict2
{'Analysis': 29, 'Machine Learning': 30}

>>> dict2['Analysis'] -2
27
```

In this case, we did not overwrite the value with the new one. Let's add 1 to 29. To do this, we need to use the following notation:

```
>>> dict2['Analysis'] = dict2['Analysis'] + 1
>>> dict2
{'Analysis': 30, 'Machine Learning': 30}
```

We can also use assignment operators presented in Chapter 2. In this case, we can subtract 2 from 30:

```
>>> dict2['Analysis'] -= 2
>>> dict2
{'Analysis': 28, 'Machine Learning': 30}
```

We can also create an empty dictionary and fill it:

```
>>> dict3 = {}
>>> dict3['key1'] = ['value1']
>>> dict3
{'key1': ['value1']}
>>> dict3['key2'] = ['value2']
>>> dict3
{'key2': ['value2'], 'key1': ['value1']}
```

One of the properties of dictionaries is called *nesting*. With nesting, we insert one dictionary into another dictionary:

```
>>> dict4 = {'key1': { 'nested1': { 'subnested1':'value1'}}}
```

At this point to get the *value* value, we have to subset like this:

```
>>> dict4['key1']['nested1']['subnested1']
'value1'
```

Sets

Sets are another Python structure. They are unordered, unduplicated items containers. They are also immutable and support typical set operations, such as union, intersection, and difference.

```
# we create a set

>>> set1 = {2, 5, 7, 9, 15}

# check its type of structure

>>> type(set1)
<type 'set'>

# and check its length

>>> len(set1)
5
```

Sets do not support indexing:

```
>>> set1[2]

Traceback (most recent call last):
  File "<stdin>", line 1, in <module>
TypeError: 'set' object does not support indexing
```

But, they do tell us whether an item is within the set:

```
>>> 9 in set1
True

>>> 17 in set1
False
```

We can also create an empty set:

```
>>> set2 = set()
```

```
>>> type(set2)
<type 'set'>
```

To fill it, we use the .add method:

```
>>> set2.add(17)
```

```
>>> set2
set([17])
```

```
>>> set2.add(24)
```

```
>>> set2
set([24, 17])
```

```
>>> set2.add(36)
```

```
>>> set2
set([24, 17, 36])
```

```
type(set2)
<type 'set'>
```

```
>>> len(set2)
3
```

Let's make another set:

```
>>> set3 = {1,1,1,1,1,2,2,2,2,2,3,3,3,3,3,4,4,4,4,5}
```

```
>>> set3
set([1, 2, 3, 4, 5])
```

```
# as you can see, a set consists of unique elements
```

Strings

Strings are character sequences that are enclosed in single or double quotation marks. They are immutable objects, but they can be repeated and combined, and parts can be extracted. We write a string like this:

```
>>> string1 = "Hi!"

# and print it

>>> string1
'Hi!'

# or write it this way with the single quotes

>>> string2 = 'Hello'

>>> string2
'Hello'

# we can print a string by writing its name or using the
print() function

>>> print(string1)
Hi!
```

A string can be composed of single words, parts of sentences, or whole sentences. Be careful when using single quotes because they may create confusion, for example, with apostrophes:

```
>>> string3 = 'I'd like to code in Python'
  File "<stdin>", line 1
    string3 = 'I'd like to code in Python'
                 ^
SyntaxError: invalid syntax
```

```
>>> string4 = "I'd like to code in Python"
```

```
>>> string4
"I'd like to code in Python"
```

However, we can include quotation marks with a backslash as follows:

```
>>> haml = "Hamlet said: \"to be or not to be ...\". Oratio answered "
```

```
>>> haml
'Hamlet said: "to be or not to be ...". Oratio answered '
```

There are some control characters that could be also useful. For example, "\n" indicates a new line:

```
>>> haml2 = "Hamlet said: to be or not to be \n Oratio answered
..."
```

```
>>> print(haml2)
Hamlet said: to be or not to be
 Oratio answered ...
```

In addition, "\t" indicates a tab:

```
>>> haml3 = "Hamlet said: to be or not to be \t Oratio answered ..."
```

```
>>> print(haml3)
Hamlet said: to be or not to be    Oratio answered ...
```

Operators that can be used when referring to strings, including the concatenation operator "+":

```
>>> string1 + string2
'Hi!Hello'
```

Or repetition operator "*":

```
>>> string1*10
'Hi!Hi!Hi!Hi!Hi!Hi!Hi!Hi!Hi!Hi!'
```

We can enter three quotation marks to mark the beginning and end of a string that extends over several lines:

```
>>> string5 = """I'd
... like
... to
... code
... in Python
... """
>>> print(string5)
I'd
like
to
code
in Python
```

We verify the class of a string with the type() function:

```
>>> type(string1)
<class 'str'>
```

And check the length with the len() function:

```
>>> len(string1)
3
```

To verify the object id, we use the id() function:

```
>>> id(string1)
4321859488
```

We can also display parts of a string:

```
>>> string1[0]
'H'
```

```
>>> string2[2]
'l'

>>> string4[-1]
'n'

>>> haml[1:10]
'amlet sai'

>>> haml[5:]
't said: "to be or not to be ...". Oratio answered '

>>> haml[:10]
'Hamlet sai'

>>> haml[:-2]
'Hamlet said: "to be or not to be ...". Oratio answere'

# the following notation is used to reverse a string (or even
just a part of it)

>>> haml[::-1]
' derewsna oitarO ."... eb ot ton ro eb ot" :dias telmaH'
```

The most important functions associated with strings allow you to start, for example, an uppercase string. We can do this by using the capitalize() method:

```
>>> string6 = "let's do a little test"
>>> string6.capitalize()
"Let's do a little test"
```

Other functions allow you to put an entire string in uppercase or lowercase letters:

```
>>> string6.upper()
"LET'S DO A LITTLE TEST"

>>> string7 = string6.upper()

>>> string7
"LET'S DO A LITTLE TEST"

>>> string7.lower()
"let's do a little test"
```

The .find method, the .index method, and the .count method are used to look for one or more characters in a string:

```
>>> string7.find("TT")
13

>>> string7.index('D')
6

>>> string7.count('L')
3
```

The strip() functions removes blank spaces at the beginning and end of a string:

```
>>> string8 = "     test     "
>>> string8.strip()
'test'
```

The replace() function allows us to replace part of a string with another element:

```
>>> string9 = "Let's do some tests"

>>> string9.replace("some", "a couple of")
"Let's do a couple of tests"
```

We can verify the presence of a substring in our string like this:

```
>>> "do" in string9
True

>>> "ueioua" in string9
False
```

With the split() function, we can break a string into a list of multiple elements:

```
>>> string9.split()
["Let's", 'do', 'a', 'little', 'test']
```

The join() function allows us to group a list into a single string:

```
>>> "-".join(["03", "01", "2017"])
'03-01-2017'
```

In the previous example, a hyphen has been inserted as a separator. The following example does not include a separator. The items are thus listed consecutively:

```
>>> "".join(["a", "b", "c", "d"])
'abcd'

# in this case we insert a space

>>> " ".join(["a", "b", "c", "d"])
'a b c d'
```

Strings are subject to immutability—meaning, they cannot be modified. Even if we can always reuse a name and overwrite it with another object inside it, but it will be a different object for all it means. Let's look at an example:

```
# we create a string

>>> string1 = "a b c d e f g"

# we try to replace the first element "a" with "x"

>>> string1[0] = 'x'

Traceback (most recent call last):
  File "<stdin>", line 1, in <module>
TypeError: 'str' object does not support item assignment

# as you can see, we get an error because we can't change the
string this way
```

The % modulus operator allows advanced string formatting. The % operator is used to search in the string for elements preceded by % and replaces them with the value or values contained in the list that follows it. The % symbol must be followed by a character that indicates the type of data we are entering in the string. To print the contents of two strings, we use "%s" like this:

```
# we create a first string

>>> string1 = 'test'

# if we want to print this part of the text and merge our
string, we enter %s before closing the quotation marks and then
insert % (string)

>>> print 'my string says: %s' %(string1)
my string says: test
```

We can use a loop to scroll a string:

```
>>> for letter in string1: print(letter)
...
t
e
s
t
```

We can count the number of letters in a string:

```
>>> word = "string test"
>>> count = 0
>>> for letter in word :
...         count = count + 1
...         print(count)
1
2
3
4
5
6
7
8
9
10
11
```

Caution   Python2 and Python3 manage strings a bit differently. In Python3, in fact, print() is a function and requires parentheses.

```
# string management in Python2

>>> print 'Hello world'
Hello world

# string management in Python3

>>> print 'Hello world'
  File "<stdin>", line 1
    print 'Hello world'
                       ^
SyntaxError: Missing parentheses in call to 'print'

>>> print('Hello world')
Hello world

# to handle strings in Python2 as they are handled in Python3,
we can import the future module:

# use of future module in Python2

>>> from __future__ import print_function

>>> print('Hello world')
Hello world
```

Files

In addition to the features we examined, we also typically import files to analyze from our computer or from the Internet. Files are often structured as dataframes, but we can also import images, audio, binary, text, or other proprietary formats, such as SPSS, SAS, a database, and so on. We learn how to import the simplest formats, such as .csv, in Chapter 6.

Immutability

As mentioned, immutability is a characteristic of some Python structures: Once created, the structure cannot be modified (see Table 3-2). We can reuse a name and overwrite the structure with another object inside it, but it will be different for all intents and purposes.

Table 3-2. *Data Structures and Immutability*

Structure	Mutable
Lists	✓
Dictionaries	✓
Tuples	✗
Sets	✗
Strings	✗

Let's look at more examples. We start first with a list, which is a mutable object:

```
# we create a list
>>> list1 = ["jan", "feb", "mar", "apr"]
# we check the type of object
>>> type(list1)
<class 'list'>
```

53

```
# and check the ID of the created list

>>> id(list1)
4302269184

# we add an element

>>> list1.append("oct")

# we reprint the list

>>> list1
['jan', 'feb', 'mar', 'apr', 'oct']

# and check the ID again

>>> id(list1)
4302269184

# as you can see, the ID is identical
```

Now, let's create a tuple, which is an immutable object:

```
>>> tuple1 = (1,2,3,4)

# we check the object class

>>> type(tuple1)
<type 'tuple'>

# and verify the ID

>>> id(tuple1)
4302119432

# we then try to add an element

>>> tuple1.append(5)
```

```
Traceback (most recent call last):
  File "<stdin>", line 1, in <module>
AttributeError: 'tuple' object has no attribute 'append'
```

```
# we recreate the tuple that also contains the last object

>>> tuple1 = (1,2,3,4,5)

# and print its contents

>>> tuple1
(1, 2, 3, 4, 5)

# and verify the ID

>>> id(tuple1)
4301382000
```

```
# as seen, we did not overwrite the first object; we created a
second object with the same name (the first tuple1 object is no
longer available)
```

Last, let's examine some examples with strings, which are immutable:

```
# we create a string

>>> string1 = "a b c d e f g"
```

```
# and try to replace the first element "a" with "x" String1 [0]
= 'x'
```

```
Traceback (most recent call last):
  File "<stdin>", line 1, in <module>
TypeError: 'str' object does not support item assignment
```

```
# we get an error
```

Converting Formats

We can transform one structure to another quite easily with the help of some functions.

```
# let's create some objects

>>> tuple1 = (1,2,3,4,5)
>>> list1 = ["jan", "feb", "mar", "apr"]
>>> string1 = "2017"
>>> int1 = 67

# we check the type of objects

>>> type(tuple1)
<type 'tuple'>

>>> type(list1)
<type 'list'>

>>> type(string1)
<type 'str'>

>>> type(int1)
<type 'int'>
```

To convert formats, we use the following functions:

```
# list() converts, for example, a tuple to a list

>>> convt1 = list(tuple1)

# it is necessary to save the result to a new object; let's do
it again and recheck the type

>>> type(convt1)
<type 'list'>
```

```
# from list to tuple

>>> conv_to_list = tuple(list1)

>>> type(conv_to_list)
<type 'tuple'>

# from string to integer

>>> conv_to_int = int(string1)

>>> type(conv_to_int)
<type 'int'>
```

Summary

In this chapter we learned how to create and manipulate the most important basic data structures in Python. An object-oriented programming language like Python is based on two main features: objects and actions. In this chapter we learned more about the objects; in Chapter 4, we focus on actions by creating functions.

Functions

An object-based programming language is structured around two major concepts: objects and functions. An object is everything we create in a work session using a programming language such as Python. Functions allow us to assign one or more actions to these objects. Let's learn how to create a function.

Some words about functions in Python

With Python, we basically have two types of functions:

1. The built-in functions that are part of Python and are loaded automatically when we run Python

2. The functions we can build and use (user defined)

A function is a piece of code that performs one or more operations on an object and returns an output result. Functions are especially useful when we have to do the same thing over multiple objects. We can do this without repeating the same line of code several times.

The two types of functions are also supported by those in the many libraries available for installation on Python. Whenever we need a particular function (or a package, that is a family of functions), we can install it and use it. Anaconda does not allow us to install many of the packages we need because they already exist in the suite.

If a package is not included in Anaconda, we can always install it using generic terms:

```
$ conda install package_name
```

Or, we can use pip:

```
$ pip install package_name
```

In any case, the exact wording for installing a package is always included in the official documentation of the package itself.

Some Predefined Built-in Functions

Default functions are within the **builtins** module. Although there are many, some of the most commonly used ones are **dir**, **help**, **type**, and **print**. We can display them by typing

```
>>> dir(__builtins__)
['ArithmeticError', 'AssertionError', 'AttributeError',
'BaseException', 'BlockingIOError', 'BrokenPipeError',
'BufferError', 'BytesWarning', 'ChildProcessError',
'ConnectionAbortedError', 'ConnectionError',
'ConnectionRefusedError', 'ConnectionResetError',
'DeprecationWarning', 'EOFError', 'Ellipsis',
'EnvironmentError', 'Exception', 'False', 'FileExistsError',
'FileNotFoundError', 'FloatingPointError', 'FutureWarning',
'GeneratorExit', 'IOError', 'ImportError', 'ImportWarning',
'IndentationError', 'IndexError', 'InterruptedError',
'IsADirectoryError', 'KeyError', 'KeyboardInterrupt',
'LookupError', 'MemoryError', 'NameError', 'None',
'NotADirectoryError', 'NotImplemented', 'NotImplementedError',
'OSError', 'OverflowError', 'PendingDeprecationWarning',
'PermissionError', 'ProcessLookupError', 'RecursionError',
```

```
'ReferenceError', 'ResourceWarning', 'RuntimeError',
'RuntimeWarning', 'StopAsyncIteration', 'StopIteration',
'SyntaxError', 'SyntaxWarning', 'SystemError',
'SystemExit', 'TabError', 'TimeoutError', 'True',
'TypeError', 'UnboundLocalError', 'UnicodeDecodeError',
'UnicodeEncodeError', 'UnicodeError', 'UnicodeTranslateError',
'UnicodeWarning', 'UserWarning', 'ValueError', 'Warning',
'ZeroDivisionError', '__build_class__', '__debug__', '__
doc__', '__import__', '__loader__', '__name__', '__package__',
'__spec__', 'abs', 'all', 'any', 'ascii', 'bin', 'bool',
'bytearray', 'bytes', 'callable', 'chr', 'classmethod',
'compile', 'complex', 'copyright', 'credits', 'delattr',
'dict', 'dir', 'divmod', 'enumerate', 'eval', 'exec',
'exit', 'filter', 'float', 'format', 'frozenset', 'getattr',
'globals', 'hasattr', 'hash', 'help', 'hex', 'id', 'input',
'int', 'isinstance', 'issubclass', 'iter', 'len', 'license',
'list', 'locals', 'map', 'max', 'memoryview', 'min', 'next',
'object', 'oct', 'open', 'ord', 'pow', 'print', 'property',
'quit', 'range', 'repr', 'reversed', 'round', 'set', 'setattr',
'slice', 'sorted', 'staticmethod', 'str', 'sum', 'super',
'tuple', 'type', 'vars', 'zip']
```

The dir() function is important because it allows us to display a list of the attributes or methods of the objects we insert inside it. For example,

```
>>> test1 = ["object1", "object2", "object3", "object4",
"object5"]
>>> dir(test1)
['__add__', '__class__', '__contains__', '__delattr__', '__
delitem__', '__delslice__', '__doc__', '__eq__', '__format__',
'__ge__', '__getattribute__', '__getitem__', '__getslice__',
'__gt__', '__hash__', '__iadd__', '__imul__', '__init__',
```

```
'__iter__', '__le__', '__len__', '__lt__', '__mul__', '__
ne__', '__new__', '__reduce__', '__reduce_ex__', '__repr__',
'__reversed__', '__rmul__', '__setattr__', '__setitem__',
'__setslice__', '__sizeof__', '__str__', '__subclasshook__',
'append', 'count', 'extend', 'index', 'insert', 'pop',
'remove', 'reverse', 'sort']
```

Attributes or methods are nothing more than actions we can take on that particular object, such as adding an item to a list, as we saw in Chapter 3:

```
>>> test1.append("pippo")
>>> test1
['object1', 'object2', 'object3', 'object4', 'object5',
'pippo']
```

We can use the type() function, which shows the type of object inserted inside it.

```
>>> type(test1)
<type 'list'>
```

It is important to remember that when bracketing an object (such as a list, tuple, dictionary, and so on) using the dir() function, we get a list of actions we can assign to that particular object.

When we work with packages and functions written by other data scientists, it is useful to obtain information about their functions and their parameters. Let's see how to do this.

Obtain Function Information

Within a function, we can find all the parameters specific to that function. To get information about a function and its parameters, type

```
>>> help(print)
Help on built-in function print in module builtins:
print(...)
    print(value, ..., sep=' ', end='\n', file=sys.stdout,
    flush=False)
    Prints the values to a stream, or to sys.stdout by default.
    Optional keyword arguments:
    file:  a file-like object (stream); defaults to the current
    sys.stdout.
    sep:   string inserted between values, default a space.
    end:   string appended after the last value, default a
    newline.
    flush: whether to forcibly flush the stream.
# the help() function is only available for Python3
```

Thus, we get a series of information about that function. To quit, press q. We can also get help regarding a particular method:

```
>>> help(test1.append)
Help on built-in function append:
append(...)
    L.append(object) -- append object to end
```

You can find the built-in functions for Python 2.7 at https:// docs.python.org/2/library/functions.html. You can find the built-in features for version 3 at https://docs.python.org/3/library/ functions.html.

If you are using Jupyter, you can display the methods by pressing the Tab key. Press Shift+Tab to display the parameters of a function (Figures 4-1 and 4-2).

```
data.
data.DataReader
data.EdgarIndexReader
data.EnigmaReader
data.EurostatReader
data.FamaFrenchReader
data.FredReader
data.get_components_yahoo
data.get_data_enigma
data.get_data_famafrench
data.get_data_fred
```

Figure 4-1. *Methods in Jupyter 1*

```
data.DataReader |
                                                              ^ + x
Signature: data.DataReader(name, data_source=None, start=None, end=None,
retry_count=3, pause=0.001, session=None, access_key=None)
Docstring:
Imports data from a number of online sources.
```

Figure 4-2. *Parameters of a function in Jupyter 2*

When using Spyder, the information in Figures 4-3 and 4-4 appears automatically.

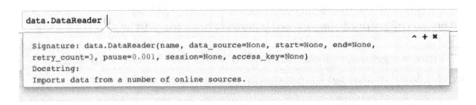

Figure 4-3. *Methods in Spyder 1*

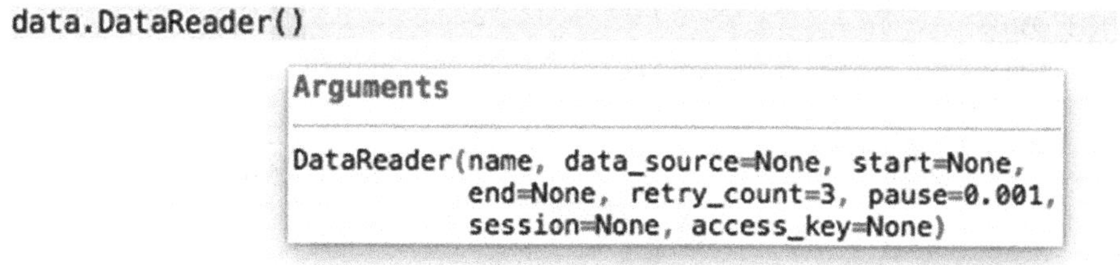

```
data.DataReader()
```

Arguments

```
DataReader(name, data_source=None, start=None,
           end=None, retry_count=3, pause=0.001,
           session=None, access_key=None)
```

Figure 4-4. *Parameters of a function in Spyder 1*

Create Your Own Functions

In addition to using the default features or importing them from other libraries, we can also create our own functions. As mentioned, functions are pieces of code that tell Python how to do something. A function has three parts: name, parameters, and body (Figure 4-5). The statement that allows us to create a function is def:

```
>>> def goal_fun(x):
...         "'(x) -> y
...         here we will write the documentation of the
            function, then
        what the function performs
...
        "'

...         return(x+y)
```

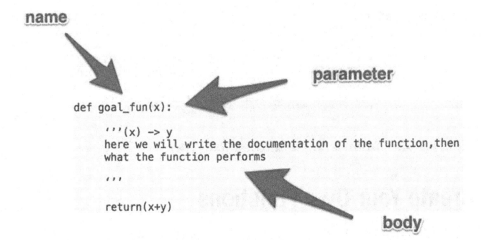

Figure 4-5. *How to write a function*

Let's create a function that sums the number 5 to any x value:

```
>>> def sum1(x):
...          "'sum x to 5
...          ""
...          return(x+5)
>>> sum1(10)
15

>>> sum1(130)
135
```

In this function, we entered one parameter, but we can enter more than one:

```
>>> def mult_xy(x, y):
...       "'multiply x and y
...       ""
...       return(x*y)
>>> mult_xy(5,6)
30
```

66

To help us see the path taken by one of our functions, we can use online tools such as Python Tutor (http://pythontutor.com/).

Save and run Your Own Modules and Files

We've seen how to create .py scripts and put them in a work directory, which we can find by importing the **os** module and typing the following:

```
>>> import os
>>> os.getcwd()
```

We can create a file from any text editor, which we must rename so it includes .py:

example_script.py

After the script is placed in the work directory, we run it by typing

```
# type on the computer terminal
$ python example_script.py
```

If the script is not in the work directory, we need to change the directory from the computer terminal:

```
# type on the terminal
$ cd directory_address
```

For instance,

```
# type on the terminal
$ cd /~/Downloads
```

After the directory has changed, proceed as we did earlier:

```
$ python example_script.py
```

The Python shell is convenient for testing on the fly, but for a complex script, it is always better to write it using an editor and then run it that way, or copy the script and run it in the Python shell.

Summary

Writing functions is a very important task for a data scientist. Languages such as R have a large number of packages and functions for every statistical need. With Python, however, we often need to write our functions, as detailed in this chapter. In Chapter 5, we look at more tools that we can use to build a useful function.

CHAPTER 5

Conditional Instructions and Writing Functions

In this chapter we explore the ways to create a function and a loop in Python. We may need to create a loop to iterate an action over a list, or to create a function to extract some cases from a dataset. Writing functions is very important to automating data analysis.

Conditional Instructions

Conditional instructions are structures used to manage conditions when we create a function. Depending on certain values or the results of an operation, we implement different actions on our data using the following conditional instructions.

- **if**
- **elif**
- **else**

© Valentina Porcu 2018
V. Porcu, *Python for Data Mining Quick Syntax Reference*,
https://doi.org/10.1007/978-1-4842-4113-4_5

These structures, along with those for creating loops, are indispensable for creating features that allow us to create special instructions, such as performing recursive operations on multiple rows of a dataset, establishing conditions, and so on.

if

Let's look at some examples of the uses of **if**:

```
>>> x = 5
>>> y = 7
>>> if x < y:
...         print("x is less than y")
>>> x is less than y
```

Now let's create objects and set a condition. The previous condition is fulfilled and the required sentence is printed.

```
>>> z = 700
>>> h = 20
>>> if h > z:
...         print("h is bigger than z")
```

In this second case, the condition is not fulfilled, therefore nothing is printed.

if + else

if can also work with **else** to give us more flexibility. For instance,

```
>>> z = 700
>>> h = 20
```

```
>>> if h > z:
...        print("h is bigger than z")
...else:
...        print("h is not bigger than z")

h is not bigger than z
```

elif

We reach maximum flexibility for if by using **elif,** which establishes intermediate conditions in a function. For instance, let's create a program that asks us what our score was. We then fit this into a scoring class, which we call *result*. Let us proceed as follows using Python2 (we might need to enter the encoding):

```
#!/usr/bin/env python
# -*- coding: utf-8 -*-

>>> print "Enter your score: "
>>> mark = int(raw_input())
>>> if 90 <= mark <= 100:
...     output = "A"
...elif 80 <= mark <= 89:
...     output = "B"
...elif 70 <= mark <= 79:
...     output = "C"
...elif 60 <= mark <= 69:
...     output = "D"
...elif mark <= 59:
...     output = "F"
...else:
...     print "I don't understand, try again"
...print "Your result is " + output
```

We can save the script (I named it mark2, and you can find it in the code folder or create by yourself by opening a .txt file and renaming it with a .py extension) and proceed by running it from Jupyter as shown in Figure 5-1.

```
In [1]: print "Enter your score: "
        mark = int(raw_input())
        if 90 <= mark <= 100:
            output = "A"
        elif 80 <= mark <= 89:
            output = "B"
        elif 70 <= mark <= 79:
            output = "C"
        elif 60 <= mark <= 69:
            output = "D"
        elif mark <= 59:
            output = "F"
        else:
            print "I don't understand, try again"
        print "Your result is " + output

        Enter your score:
        70
        Your result is C
```

Figure 5-1. *Running mark2 in Jupyter*

Caution   Python2 and Python3 manage user input differently.

In Python 3, we simply use input() instead of raw_input():

```
>>> print("Enter your score: ")
>>> score = int(input())
>>> if 90 <= score <= 100:
...     output = "A"
...elif 80 <= score <= 89:
...     output = "B"
...elif 70 <= score <= 79:
...     output = "C"
```

```
...elif 60 <= score <= 69:
...    output = "D"
...elif score <= 59:
...    output = "F"
...else:
...    print("I don't understand, try again")
...print("Your result is " + output)
```

Enter your score:

80

Your result is B

Loops

Loops identify structures that allow you to repeat a certain portion of code, for a number of times or under certain conditions. The most important instructions in Python that allow you to tweak actions are

- **for**
- **while**
- **continue and break**

for

The Python instruction **for** allows the definition of iterations. **for** is an iterator, so it is able to go through a sequence and perform actions on it, or perform operations. The format of **for** instructions is as follows:

for item in object:
```
    run action on item
```

For instance,

```
# we create an object (in this case, a tuple) and print every
element

>>> x = (1,2,3,4,5,6,7)

# check the type of object

type(x)
<type 'tuple'>

>>> for n in x:
...        print(n)

1
2
3
4
5
6
7

# we create an object and, for each element of the object, we
print a sentence together with the element

>>> x = (1,2,3,4,5,6,7)

>>> for n in x:
...     print("this is the number", n)

("this is the number", 1)
("this is the number", 2)
("this is the number", 3)
("this is the number", 4)
("this is the number", 5)
```

```
("this is the number", 6)
("this is the number", 7)

# we can also print the elements of a string

>>> string1 = "example"

>>> for s in string1:
...         print(s)

e
x
a
m
p
l
e

>>> for s in string1:
...         print(s.capitalize())

E
X
A
M
P
L
E

>>> for s in string1:
...         print(s*5)

eeeee
xxxxx
aaaaa
```

```
mmmmm
ppppp
11111
eeeee
```

What happens if, for example, we have a list that contains sublists and we want to print some of its elements? Let's look at an example:

```
# we create a list containing pairs of elements
>>> list1 = [(5,7), (9,2), (2,3), (14,27)]
>>> list1
[(5, 7), (9, 2), (2, 3), (14, 27)]
```

We want to print only the first element of each of the pairs, as shown in Figure 5-2.

```
>>> for (el1, el2) in list1:
...      print el1

5
9
2
14
```

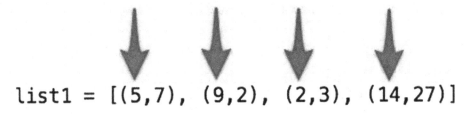

Figure 5-2. *Print the first element only*

If we want to, we can also carry out operations on couples—for example,

```
>>> for (el1, el2) in list1:
...         print el1+el2

12
11
5
41

# In Python3, we add parentheses to the function print()
>>> for (el1, el2) in list1:
...         print(el1+el2)

12
11
5
41
```

As for dictionaries, we can proceed as follows:

```
>>> dict1 = {"k1":1, "k2":2, "k3":3}

>>> for key, value in dict1.items():
...     print("the key value " + str( key ) + " is " +
        str( ...value ))

the key value k3 is 3
the key value k2 is 2
the key value k1 is 1
```

Clearly, we can also print a single key or the only value:

```
# python2

>>> for key, value in dict1.items(): print key

k3
k2
k1

>>> for key, value in dict1.items(): print value

3
2
1

# python3

>>> for key, value in dict1.items(): print(key)

k3
k2
k1

>>> for key, value in dict1.items(): print(value)

3
2
1
```

while

The instruction **while** executes actions if a certain condition is met. It is used in cases when we do not know with certainty how many times an iteration has to be processed, so we run it until it satisfies a specific condition.

```
# Python2

>>> x = 1
>>> while x < 5:
...     print x
...     x = x+1

1
2
3
4

# Python3

>>> x = 1
>>> while x < 5:
...     print(x)
...     x = x+1
```

With the **while** instruction, it is necessary to be careful not to start an infinite loop, which would then require you to force the program to close. For instance, if we set this type of loop we will get a list of "1" until we stop the execution:

```
# nb: don't run

>>> x = 1
>>> while x < 2:
...         print x
```

Let's look at another example of **while**:

```
# Python2

>>> y = 1
>>> while y < 10:
...     print "the y value is " ,y
...     y = y +1
```

```
the y value is  1
the y value is  2
the y value is  3
the y value is  4
the y value is  5
the y value is  6
the y value is  7
the y value is  8
the y value is  9
```

```
# in this case, we print a sentence with each of the y values,
which increases at each new step
```

continue and break

continue and **break** are two instructions that allow you to end a cycle or to continue passing to the next iteration. Let's look at an example of **continue**:

```
# we create a list

>>> list1 = ['item1', 'item2', 'item3', 'cat', 'item4', 'item5']

# we create a for loop that prints each items in the list

>>> for item in list1:
...     if item in list1:
...         if item == 'cat':
...             continue
...     print(item)

item1
item2
item3
item4
item5
```

the for loop must skip the element that is unlike the others;
to get this result, we use continue, thus "skipping"' the element

break works in a similar way, although unlike **continue**, it interrupts the **for** cycle:

```
>>> for item in list1:
...     if item in list1:
...         if item == 'cat':
...             break
...     print(item)
item1
item2
item3
```

range()

As we have seen for lists, range() is a function in Python2 (in Python3, it is a built-in method) and not a conditional instruction. This function allows you to create lists bounded by an upper limit and a lower limit (a range).

```
# Python2

# we create a list of numbers from 1 to 49 (included)

>>> list = range(1, 50)

>>> list
[1, 2, 3, 4, 5, 6, 7, 8, 9, 10, 11, 12, 13, 14, 15, 16, 17, 18,
19, 20, 21, 22, 23, 24, 25, 26, 27, 28, 29, 30, 31, 32, 33, 34,
35, 36, 37, 38, 39, 40, 41, 42, 43, 44, 45, 46, 47, 48, 49]

# check the object type

>>> type(list)
<type 'list'>
```

```
# we create a list of numbers from 30 to 44 (included)
>>> list2 = range(30, 45)
>>> list2
[30, 31, 32, 33, 34, 35, 36, 37, 38, 39, 40, 41, 42, 43, 44]
# if we do not specify the lower limit, but only the upper
limit, the list starts at 0:
>>> list3 = range(14)

>>> list3
[0, 1, 2, 3, 4, 5, 6, 7, 8, 9, 10, 11, 12, 13]
```

We can also add the range we want to include, for example, of three elements, as the third argument: so the first parameter will be the number from to start with (20), the second the end of the list (40) and the third will be the step (3)

```
>>> list4 = range(20, 40, 3)

>>> list4
[20, 23, 26, 29, 32, 35, 38]
# this way, we can, for example, create lists for odd or even numbers
>>> range(0,10,2)
[0, 2, 4, 6, 8]

>>> range(1,10,2)
[1, 3, 5, 7, 9]
```

We can also create two objects and require a range between the two objects. For example,

```
>>> x = 17
>>> y = 39
```

```
>>> range(x, y)
[17, 18, 19, 20, 21, 22, 23, 24, 25, 26, 27, 28, 29, 30, 31,
32, 33, 34, 35, 36, 37, 38]
```

Also, we can incorporate lists created with range()within another loop. For example,

```
>>> for el in range(1,10): print el

1
2
3
4
5
6
7
8
9
# for example, instead of passing an object in the for loop, we
pass a list created with range()
```

Caution   In Python2 and Python3, range() is used differently.

To get a list of numbers in Python3, we write

```
>>> list(range(10))
[0, 1, 2, 3, 4, 5, 6, 7, 8, 9]
>>> list(range(2,10))
[2, 3, 4, 5, 6, 7, 8, 9]
# if we use the same name as Python2, we get this result in
Python3
>>> range(10)
>>> range(0, 10)
```

Extend Functions with Conditional Instructions

The conditional and loop instructions we just studied allow us to extend our capabilities in writing functions. Let's look at an example:

```
# we create an example function

>>> def ex1(x):
...         y = 0
...         while(y < x):
...             print('Add one!')
...             y = y + 1
...         return('Stop now!')

>>> ex1(0)
'Stop now!'

ex1(5)
Add one!
Add one!
Add one!
Add one!
Add one!
'Stop now!'
```

map() and filter() Functions

The map() function takes two objects, a function, and a sequence of data, and applies the function to the data sequence. Here is an example:

```
# map()
# Python2
```

```
# we create a function that squares a number

>>> def square(x):
...         return x*x

# check if the function does what it is supposed to do

>>> square(9)
81

# now let us create a list of numbers

>>> num = [2, 5, 7, 10, 15]
```

we apply the map() function to our list to square all list numbers with only one operation

```
>>> map(square, num)
[4, 25, 49, 100, 225]

# Python3
```

we use the list() function to get the desired result, as we did with range()

```
>>> list(map(square, num))
```

The filter() function applies a function to an object and returns the results that meet a particular criterion. Let's create a second list of numbers:

```
>>> num2 = range(1,15)

>>> num2
[1, 2, 3, 4, 5, 6, 7, 8, 9, 10, 11, 12, 13, 14]
```

we create a function that allows us to distinguish even numbers

```
>>> def even(x):
...         if x % 2 == 0:
...                 return True
...         else:
...                 return False
```

```
# we apply the filter() function to the function even() and a
list of numbers, like num2
```

```
>>> filter(even, num2)
[2, 4, 6, 8, 10, 12, 14]
```

The lambda Function

The lambda function is a Python construct with a particular syntax that simplify a function construction allowing us to reduce one function in one line, thus making the code more simple but less powerful than creating a function with the construct def. For instance, to create a function that squares a number, we did this:

```
# we create a function that squares a number
```

```
>>> def square(x):
...        return x*x
```

```
# we check if the function does what it is supposed to do
```

```
>>> square(9)
81
```

Now we use the lambda function to create the same function:

```
>>> sq2 = lambda x : x*x
```

```
>>> sq2(127)
16129
```

86

As you can see, we can create a similar function in a single line.

Let's return to a list of numbers:

```
>>> num =[2, 5, 7, 10, 15]
```

```
# now we use the second function we created to apply the
function to the entire list
```

```
>>> map(sq2, num)
[4, 25, 49, 100, 225]
```

```
# it is not necessary to create a function; we can integrate
the lambda function directly within the process. In this case,
if we want to use the same lambda function over there, we will
rewrite it instead than recalling it by its name
```

```
>>> map(lambda x: x*x, num)
[4, 25, 49, 100, 225]
```

Scope

When writing a function, we can define an object within the function itself or relate to an object created externally by the function. In the first case, we speak of a *global variable*; in the second case, a *local variable*. With respect to a function, we therefore have three elements:

1. **Formal parameters:** the arguments present in the definition of the function

2. **Local variables:** which are defined by evaluating expressions in the body of the function and have visibility only within the function itself

3. **Global variables:** which do not belong to either the first or the second group, but are looked for outside the function

So far we have worked with global variables, defining them first and then applying various operations through some functions. In the following case, however, let's define a global test variable external to the function and a local test function within the function. The function calls the local variable.

```
>>> test = "hello"

>>>    def fun1():
...        test = "hi"
...        print(test)

>>> fun1()
hi
```

In this next case, however, let's refer to a variable external to the function itself. In so doing, the calculation is performed correctly:

```
num1 = 5

def fun2():
    global num1
    num2 = num1 + 1
    print(num2)

fun2(num1)
6
```

Summary

In this chapter we examined conditional instructions, which are structures used to manage conditions when we create functions. We also looked at extending our functions.

CHAPTER 6

Other Basic Concepts

In this chapter, we learn about some important programming concepts (such as modules and methods), list comprehension and class creation, regular expressions, and management of errors and exceptions.

Object-oriented Programming

As mentioned, Python is an object-oriented programming language, so let's look at some basic concepts of object-oriented programming. Some important concepts are

- Objects
- Classes
- Inheritance

More on Objects

Objects are all the data structures we have created, from smaller ones containing only a single number to larger ones containing large datasets. We apply objects to operations using features or methods that are preinstalled in libraries, that we 'import, or that we create.

© Valentina Porcu 2018
V. Porcu, *Python for Data Mining Quick Syntax Reference*,
https://doi.org/10.1007/978-1-4842-4113-4_6

Classes

As far as the class concept is concerned, this is a new topic for us in this book; one we have not yet examined. Python allows us to create structures linked to our needs through the class concept. Classes are abstract representations of an object that we can fill with real instances from time to time. When a new class is defined, we can define instances of that class. A class is defined by its own characteristics. For example, if we create a Dog class, we can stipulate that it be defined by features such as the shape of the head, muzzle, hair type (short or long), and so on. A Book class could include features such as book genre, the number of pages, the main topic, the type of cover, ISBN and so on.

Inheritance

Another concept behind object-oriented programming is inheritance. It is possible to create new classes from existing classes. New classes inherit the characteristics of the original classes, but they can extend them with new features. Inheritance is convenient, because it allows us to extend old classes without having to change them. Inheritance can be single and multiple. Through single inheritance, a subclass can inherit member data and methods from one existing class; in the case of multiple inheritance, a subclass can inherit characteristics from more than one existing class.

Let's look at a simple example that creates a Cat class.

```
>>> class Cat:
... def __init__(self, name, color, age, race):
...     self.name = name
...     self.color = color
...     self.age = age
...     self.race = race
```

Cat is defined by name, color, age, and breed. We can create a cat instance:

```
>>> cat1 = Cat("Fuffy", "white", 3, "tabby")
```

which let's see its features:

```
>>> print(cat1.name)
Fuffy
>>> print(cat1.color)
white
```

We can modify the class by adding methods:

```
>>> class Cat:
...   def __init__(self, name, color, age, race):
...       self.name = name
...       self.color = color
...       self.age = age
...       self.race = race

...   def cry(self):
...       print("meow")
...   def purr(self):
...       print("purr")
# we create a cat instance with this new class
>>> cat2 = Cat("Candy", "Red", "6", "Balinese")
```

Now we can query the instance not only on the basis of the features, as we did in the first example,

```
>>> print(cat2.age)
6
```

but also by using the methods:

```
>>> cat2.cry()
meow

>>> cat2.purr()
purr
```

Let's create a subclass for tabby, a cat breed, so it inherits its features.

```
>>> class tabby(Cat):
...     def character(self):
...         print("warm")
# we create a tabby instance
>>> tabby1 = tabby("Pallina", "black", 4, "tabby")
>>> tabby1.purr()
purr

>>> tabby1.character()
warm
```

Thus we can query the instance not only with regard to the characteristics of the tabby subclass but also with regard to the existing Cat class.

Modules

Modules are collections of functions that are generally related to a given topic (graphics, data analysis). Forms can belong to one of the following categories:

- Python modules
- Precompiled modules
- Built-in modules

To use a module, we must first import it by using the import instruction:

import ...

For instance,

```
import numpy
```

Using ""import,"" we imported the entire module.
However, we can import part of a module:

from ... import ...

For example,

```
from matplotlib import cm
```

To simplify and speed up the writing of code, we can import a form with another name:

```
import numpy as np
```

We can then check our modules with the following command:

```
help('modules')
```

After a module is imported—say, in Jupyter—we can access the help section by displaying all the methods of that particular module by pressing the Tab key (Figure 6-1).

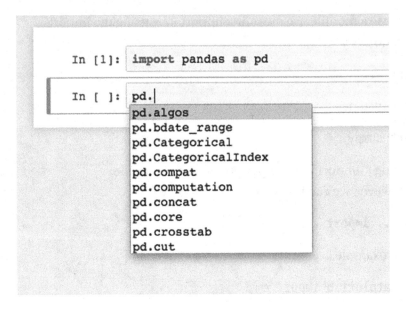

Figure 6-1. *pandas' methods on Jupyter*

In addition, we can import very specific elements of a module. For instance:

```
from math import sqrt
```

In this case, we are not importing the entire module—only the function for the square root. In this case, we don't need to call sqrt as a method of the **math** module. We can call the square root directly:

```
>>> sqrt(9)
3.0
```

If we import the entire module, we must specify sqrt as the math module method:

```
>>> import math
>>> math.sqrt(12)
3.4641016151377544
```

Some programming languages, such as Anaconda, install a whole series of modules and packages automatically (Figure 6-2). However, if we need to install a particular module, we can do it from the computer terminal by typing:

```
$ pip install nome_modulo
```

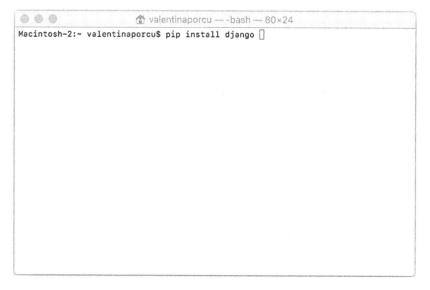

Figure 6-2. *Installing a package on the computer*

As you can see in Figure 6-2, we are on the terminal, not in a Python window, because as the ">>>" symbol is missing at the prompt. The "$" symbol indicates that the terminal is being used.

Depending on the package we are installing (for example, from GitHub), we may find different instructions for installation in the package documentation itself. Packages are a collection of modules, often on the same subject. For instance, SciPy and NumPy contain dozens of data analysis modules.

Methods

In Python, everything is an object and, depending on the type to which it belongs, different methods (or functions) can be applied to each object. Methods are sort of like functions, but they are related to particular classes. This means that lists have their own methods, tuples have different methods, and so on. Each method performs an operation on an object, similar to a function.

Depending on the tool we use to program or for our Python data analyses, we may have some suggestions on methods associated with a particular object. This is the case when using Spyder and Jupiter. Figure 6-3 displays an example of methods for an object using Spyder; Figure 6-4 shows the methods using Jupyter.

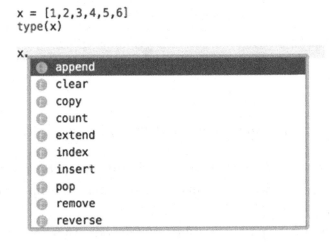

```
x = [1,2,3,4,5,6]
type(x)
```

```
x.
    append
    clear
    copy
    count
    extend
    index
    insert
    pop
    remove
    reverse
```

Figure 6-3. *Example on Spyder*

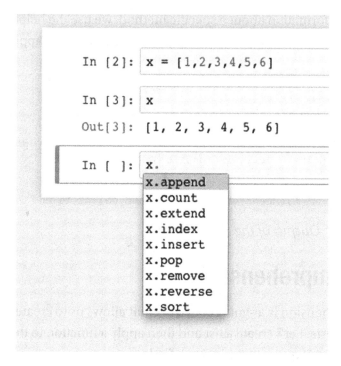

Figure 6-4. *Example on Jupyter*

An example of a function is print(); an example of a method is .upper. Let's look at an example:

```
>>> string1 = "this is a string"
>>> print(string1)
this is a string

>>> string1.upper()
"THIS IS A STRING"
```

To get information about a specific method, we use the help() function. For example, we can create a list and activate help with the .append method:

```
>>> x = [1,2,3,4,5,6]
>>> help(x.append)
```

```
Help on built-in function append:

append(...)
        L.append(object) -- append object to end
```

Figure 6-5. *Output of the code above*

List Comprehension

List comprehension is a syntax construct that allows us to create new lists from other lists. Let's create a list and then apply a function to the elements of another list. We can do this using a **for** loop:

```
# we create a number list
>>> numbers = [12, 23, 34, 57, 89, 97]
# in Python2
# using the for loop we add the number 10 to each item on the
list
>>> for i in range(len(numbers)):
...            numbers[i] = numbers[i] + 10
>>> numbers
[22, 33, 44, 67, 99, 107]

# we can create a new list by adding a certain number even
through the list comprehension
# we will overwrite the first list, number, with a new one
where every item is added to 10 using Python3
>>> numbers = [number + 10 for number in numbers]
```

```
[32, 43, 54, 77, 109, 117]
# the way we call each element of the list is random
>>> numbers = [n + 10 for n in numbers]
>>> numbers
[42, 53, 64, 87, 119, 127]
```

Thus, list comprehension allows us to simplify an iteration that turns our list into a new one. We can apply list comprehension to a list of numbers, as we just saw, we can also apply it to strings. For example, we can iterate the operation and turn everything in it into uppercase type:

```
>>> strings = ['this', 'is' ,'a', 'string']
>>> strings2 = [string.upper() for string in strings]

>>> strings2
['THIS', 'IS', 'A', 'STRING']
```

Regular Expressions

Imagine trying to search for a word in a text document. When we do the search, it returns all instances of the word we are looking for. However, if we do this type of search, we may skip some occurrences—for example, if they start with an uppercase letter or if they are followed or preceded by punctuation. If the word we are looking for also exists within another word, we must also consider the empty spaces of the word itself to find it.

The use of regular expressions, also known as *regex*, makes it much easier to identify all these options. Regular expressions are patterns that allow us to describe generic strings, which make this type of search more elaborate. They are very useful for searching, replacing, or extracting strings or substrings. Regular expressions can, for example, be used to extract dates, e-mail addresses, and physical mail addresses, because they do not just extract the single e-mail address. For example, we can insert an e-mail address in a search box with a structure similar to address@email.com, which identifies the structure of an e-mail address and extracts multiple e-mail addresses from the same document according to its structure.

In Python, regular expressions are handled via the module **re**, which we can import:

```
>>> import re
# we create a sample string
>>> str1 = "Try searching for a word using regular expressions
and the Python module kernel"
```

We search for an occurrence using re.search():

```
>>> re.search('word', str1)
<_sre.SRE_Match object at 0x10280bed0>
# this result tells us that the word we were looking for is
present in the string
# to get the same result in its simplest form, save the
previous line of code in an object and query it with the bool()
function:
>>> exre1 = re.search('word', str1)
>>> bool(exre1)
True
```

We can search for an occurrence by using .findall():

```
>>> re.findall('Try', str1)
['Try']

# if the item is not present, we receive an empty list
>>> re.findall('some', str1)
[]

# this function is case sensitive, so "'some"' is different
from "'Some"'
```

We can also divide a string into the elements that comprise it:

```
>>> re.split(' ', str1)
['Try', 'searching', 'for', 'a', 'word', 'using', 'regular',
'expressions', 'and', 'the', 'Python', 'module', 'kernel']
# in the previous code, we used a space as a splitting element;
next we split the string into words using the conjunction '
and ' (and insert spaces on either side to avoid searching for
internal recurrences of a word)
>>> re.split(' and ', str1)
['Try searching for a word using regular expressions', 'the
Python module kernel']
# we get the previous result by dividing the string in two
according to the position of the conjunction "'and'"
```

Regular expression symbols (such as *, +, and ?) allow us to search for a character one or more times, or followed by other letters:

```
>>> re.findall('ea*', str1)
['ea', 'e', 'e', 'e', 'e', 'e', 'e', 'e']
>>> re.findall('ea+', str1)
['ea']

>>> re.findall('ea?', str1)
['ea', 'e', 'e', 'e', 'e', 'e', 'e', 'e']
>>> re.findall('ea+?', str1)
['ea']

# we can, for instance, extract all the words with a capital
letter
>>> re.findall('[A-Z][a-z]*', str1)
['Try', 'Python']
```

```
# or all the words in lowercase letters
>>> re.findall('[a-z]*', str1)

['',
 'ry',
 '',
 'searching',
 '',
 'for',
 '',
 'a',
 '',
 'word',
 '',
 'using',
 '',
 'regular',
 '',
 'expressions',
 '',
 'and',
 '',
 'the',
 '',
 '',
 'ython',
 '',
 'module',
 '',
 'kernel',
 '']
```

```
# or all uppercase or lowercase words
>>> re.findall('[a-z]*', str1, re.IGNORECASE)
['Try',
 '',
 'searching',
 '',
 'for',
 '',
 'a',
 '',
 'word',
 '',
 'using',
 '',
 'regular',
 '',
 'expressions',
...
```

```
# or we can do this
>>> re.findall('[^.\-! ]+', str1) >>> ['Try', 'searching',
'for', 'a', 'word', 'using', 'regular', 'expressions', 'and',
'the', 'Python', 'module', 'kernel']
# \d find the numbers
>>> str2 = "We are going to meet today at 14:15"
>>> re.findall('\d', str2)
# we can use other symbols to help us target the extraction
more explicitly
>>> re.findall('\d+', str2)
['14', '15']
```

```
# for example, let's look for all the words that include p
>>> re.findall(r'[p]\S*', str1)
```

```
['pressions']
# or for the letter 'p' in either lowercase or uppercase letters
>>> re.findall(r'[p]\S*', str1, re.IGNORECASE)
['pressions', 'Python']

# or we can extract e-mails from a string
>>> str3 = "my email is mail@mail.com, my second email is
test2@ex.com"
>>> re.findall("[\w\.-]+@[\w\.-]+", str3, re.IGNORECASE)
['mail@mail.com', 'test2@ex.com']
```

We can perform a regular expression test by using tools on the Web, such as http://pythex.org. (http://pythex.org/)More information about regular expressions can be found at https://docs.python.org/2/library/re.html for the **re** module. Table 6-1 lists the various symbols of regular expressions.

Table 6-1. *Symbols and Regular Expressions*

Symbol	Description
\\d	Digit, 0, 1, 2, . . . 9
\\D	Not digit
\\s	Space
\\S	Not space
\\w	Word
\\W	Not word
\\t	Tab
\\n	New line
^	Beginning of the string
$	End of the string

(continued)

Table 6-1. (*continued*)

Symbol	Description
\	Escape special characters—for example,\\ is "\", \+ is "+"
|	Alternation match—for example /(e|d)n/ is ""en"" and ""dn""
*	Any character, except \n or a line terminator
[ab]	a or b
[^ab]	Any character except a and b
[0-9]	All digits
[A-Z]	All uppercase letters from A to Z
[a-z]	All lowercase letters from a to z
[A-z]	All uppercase and lowercase letters from a to z
i+	i at least one time
i*	i zero or more times
i?	i zero or one time
i{n}	i that occurs n times in sequence
i{n1,n2}	i that occurs n1 - n2 times in sequence
i{n1,n2}?	Nongreedy match, see previous example
i{n,}	i occurs ≥n times
[:alnum:]	Alphanumerical characters: [:alpha:] and [:digit:]
[:alpha:]	Alphabetical characters: [:lower:] and [:upper:]
[:blank:]	Blank characters, such as space and tab
[:cntrl:]	Control characters
[:digit:]	Digits: 0 1 2 3 4 5 6 7 8 9
[:graph:]	Graphical characters: [:alnum:] and [:punct:]

(*continued*)

Table 6-1. (*continued*)

Symbol	Description
[:lower:]	Lowercase letters in the current locale
[:print:]	Printable characters: [:alnum:], [:punct:] and space
[:punct:]	Punctuation characters such as ! "" # $ % & " ( ) * + , - . / : ; < = > ? @ [\ ] ^ _ ' {	} ~
[:space:]	Space characters: tab, new line, vertical tab, form feed, carriage return, space
[:upper:]	Uppercase letters in the current locale
[:xdigit:]	Hexadecimal digits: 0 1 2 3 4 5 6 7 8 9 A B C D E F a b c d e f

User Input

The input function (raw_input in Python2) is used to let us talk to a program that has to handle responses depending on the type of input (we studies this when setting functions). In Python2, we handle user input with the raw_input() function:

```
>>> name = raw_input("What is your name? ")
What is your name? Valentina
>>> print(name)
Valentina

>>> print("Nice to meet you, " + name)
Nice to meet you, Valentina
```

Caution   Python2 and Python3 handle user input differently.

In Python3, we use input() instead of raw_input():

```
>>> name = input("What is your name? ")
What is your name? Valentina
>>> print(name)
Valentina

>>> print("Nice to meet you, " + name)
Nice to meet you, Valentina
```

When input is entered in this way, it is read as a string. So, for example, if we want to enter a number, we have to write the code a bit differently. We must specify that what we are entering must be read as a number:

```
>>> num1 = input('add a number: ')
>>> num2 = input('add a second number ')
>>> print(num1 + num2)
# the result will be an integer number resulting from the
addition and depending from the number you choose:
3725
```

Numbers are not summed; they are attached, as happens with two strings. To add them, specify that the value we are entering is a number, and proceed as follows:

```
>>> num1 = int(input('enter a number: '))
>>> num2 = int(input('add a second number '))
>>> print(num1 + num2)
```

Errors and Exceptions

Errors and exceptions in Python are nothing more than abnormal or unexpected events that change the normal running of our code. An exception may be the result of invalid inputs (for example, we ask users to enter a number and they enter a letter), hardware issues, or files or objects are not found. There are three main types of errors:

1. Syntactic

2. Semantic

3. Logical

Syntax errors are mistakes we make when writing code. They are either spelling mistakes or syntax errors in the code.

```
# example of a syntax error message due to the absence of the
quotation mark at the bottom of the string
>>> print 'Hello World
  File "<stdin>", line 1
    print 'Hello World
                      ^
SyntaxError: EOL while scanning string literal
```

Errors and exceptions usually cause error messages, which we can then use to identify the error and determine whether we can remedy it by modifying the code or handling an exception.

When we expect an exception to occur (called a *handled exception*), the way to remedy it is to write suitable code. Unexpected exceptions are called *unhandled exceptions*.

To handle errors and exceptions in Python, we typically use try(), except(), and raise().

For example, let's sum two items that cannot be summed, such as a number and a string:

```
>>> 37 + 'string'

TypeError                               Traceback (most
recent call last)
<ipython-input-1-5dc2db43a4bf> in <module>()
----> 1 37 + 'string'
TypeError: unsupported operand type(s) for +: 'int' and 'str'
# clearly, the result is an error
```

Let's look at the type of error in the message: TypeError. We need to create a way to handle this error. For example, let's ask users to enter two numbers and then return the sum of the two numbers.

```
>>> try:
    ... num1 = int(input('enter a number: '))
    ... num2 = int(input('enter a second number '))
    ... print(num1 + num2)
except TypeError:
    ... print("There is something wrong! Check again!")
>>> enter a number: 37
>>> enter a second number 25
62
```

We managed the TypeError error. If users insert two numbers, they are summed correctly.

Let's see what happens if an incorrect value is entered rather than a number:

```
>>> try:
    ... num1 = int(input('enter a number: '))
    ... num2 = int(input('enter a second number '))
    ... print(num1 + num2)
```

```
except TypeError:
    ... print("There is something wrong! Check again!")
>>> enter a number: 37
>>> enter a second number string
```

```
ValueError                              Traceback (most
recent call last)
<ipython-input-16-566345f8fed9> in <module>()
      1 try:
      2     num1 = int(input('enter a number: '))
----> 3     num2 = int(input('enter a second number '))
      4     print(num1 + num2)
      5 except TypeError:
ValueError: invalid literal for int() with base 10: 'string'
```

There is a problem! We managed the TypeError error, but now there is a different error. We can handle this as follows:

```
>>> try:
    ... num1 = int(input('enter a number: '))
    ... num2 = int(input('enter a second number '))
    ... print(num1 + num2)
except TypeError:
    ... print("There is something wrong! Check again!")
except ValueError:
    ... print("There is something wrong! Check again!")
enter a number: 37
enter a second number string
There is something wrong! Check again!
```

Or, we manage the exception with Exception:

```
>>>try:
    ... num1 = int(input('enter a number: '))
    ... num2 = int(input('enter a second number '))
    ... print(num1 + num2)
except Exception:
    ... print("There is something wrong! Check again!")
>>> enter a number: 37
>>> enter a second number test
There is something wrong! Check again!
```

Exception is a class of basic errors that includes most errors. Other common types, as we just saw, are TypeError and ValueError, but there are others: AttributeError, EOFError, IOError, IndexError, KeyError, KeyboardInterrupt, NameError, StopIteration, and ZeroDivisionError.

Summary

In this chapter we studied basic concepts of programming in Python: modules and methods, list comprehension and class creation, regular expressions, and errors and exceptions. In Chapter 7, we learn about importing files.

CHAPTER 7

Importing Files

Importing a file to Python is important to learning how to manage datasets. In this chapter we examine the basics. We can import many types of data to Python: from the most canonical format (.csv) or Excel data formats, to text formats for text mining, and to binary files such as images, video and audio. First, let's look at some basic ways to import files. Sometimes the process for doing so may seem a bit tricky. The pandas package, which we examine in Chapter 8, makes importing datasets for analysis much easier.

The basic structure for importing files is as follows:

```
file1 = open("file.name", mode)
```

"mode" represents the way we open a file. The most important ways are

- Read only ("r")

- Write ("w")

- Append some text at the end of the document ("a")

- Read and write ("r+")

We use the open() function to open the file. If the file does not exist, it is created in the work directory.

```
# we open (or create) the file

>>> file1 = open("file1.txt","w")

# at the moment, the file was created in our work directory but
it is empty
```

© Valentina Porcu 2018
V. Porcu, *Python for Data Mining Quick Syntax Reference*,
https://doi.org/10.1007/978-1-4842-4113-4_7

```
# we are in write mode; we then add text this way

>>> file1.write("Add line n.1 to the file1")

# we close the file

>>> file1.flush()
>>> file1.close()

# we open the file in read mode

>>> file1 = open("file1.txt", "r")

# we create an object that contains our file and display it
with the print() function

>>> text1 = file1.read()

>>> print(text1)
Add line 1 to file1

# we close the file

>>> file1.close()

# we reopen the file in write mode

>>> file1 = open("file1.txt","w")

# if we write the new text now, the line we had written
previously is replaced

>>> file1.write("Let's replace the first line with this new one")

# we test

>>> file1.flush()
>>> file1.close()
```

```
>>> file1 = open("file1.txt", "r")

>>> text2 = file1.read()

>>> print(text2)
Let's replace the first line with this new one

>>> file1.close()

# to add text without overwriting, we open the file in append mode

>>> file1 = open("file1.txt", "a")

>>> file1.write("\n Add a second line to this new version of
the file")

>>> file1.close()

>>> file1 = open("file1.txt", "r")

>>> text3 = file1.read()

>>> print(text3)
We replace the first line with this new one
 Add a second line to this new version of the file

# we can verify the length of the text with the len function()

len(text3)
105
```

We can also create a small function that reads every line of the file:

```
>>> file1 = open("file1.txt", "r")

>>> for line in file1:
...         print(line, end = "")
```

Or proceed as follows:

```
>>> with open("file1.txt", "a") as file:
...         file1.write("this is the third line")
```

Table 7-1 provides a summary of the modes used to open a file.

Table 7-1. *Modes for Opening a File*

Mode	Description
'r'	Read only, default mode
'rb'	Read only in binary format
'r+'	Read and write
'rb+'	Read and write in binary format
'w'	Write
'wb'	Write in binary format only. Overwrites an existing file. If the file does not exist, a new one is created.
'w+'	Read and write. Overwrites an existing file. If the file does not exist, a new one is created.
'wb+'	Read and write in binary format. Overwrites an existing file. If the file does not exist, a new one is created.
'a'	Adds to an existing file without overwriting. If the file does not exist, a new one is created.
'ab'	Adds to an existing file or creates a new binary file
'a+'	Reads, adds, and overwrites a new file (or creates a new one)
'ab+'	Reads and adds in binary format; overwrites a new file or creates a new one

.csv Format

Files in .csv or .tsv format are those used most frequently in data mining. Later, we look at some packages (such as pandas) that make it easier to import and manage files. For now, however, we study some basic procedures that do not require separate installation.

```
# we import csv

import csv

# next, I generate a random csv file that I save in the work
directory and call it 'df'. The second argument, 'r', means we
are accessing the file in read mode.

csv1 = open('df', 'r')

# if we want to import a file that is not in the work
directory, we can include the entire address, for example:

csv2 = open('/Users/valentinaporcu/Desktop/df2', 'r')

# we go on reading the first file

read = csv.reader(csv1)

 for row in read:
...     print row
...
['', '0', '1', '2', '3', '4']
['0', '15.982938813888007', '96.04182101708831',
'74.68301929612825', '31.670249691004994', '50.37042800222742']

[...]
```

From the Web

The basic methods also allow us to read a file from the Web. For example:

```
# Python2

# we import csv and urllib2

import csv
import urllib2

# we create an object that contains the address

url = "https://archive.ics.uci.edu/ml/machine-learning-
databases/iris/iris.data"

# we create a connection

conn = urllib2.urlopen(url)

# we create an object containing the .csv file

file = csv.reader(conn)

we print the file

for row in file:
...     print row
...
['5.1', '3.5', '1.4', '0.2', 'Iris-setosa']
['4.9', '3.0', '1.4', '0.2', 'Iris-setosa']
['4.7', '3.2', '1.3', '0.2', 'Iris-setosa']
['4.6', '3.1', '1.5', '0.2', 'Iris-setosa']
['5.0', '3.6', '1.4', '0.2', 'Iris-setosa']
['5.4', '3.9', '1.7', '0.4', 'Iris-setosa']

[...]
```

In JSON

Let's see how to import a test file in JSON. Some JSON test files can be downloaded from https://www.jsonar.com/resources/.

```
import json

jfile = open('zips.json').read()

print(jfile)
```

```
{ "city" : "AGAWAM", "loc" : [ -72.622739, 42.070206 ], "pop" :
15338, "state" : "MA", "_id" : "01001" }
{ "city" : "CUSHMAN", "loc" : [ -72.51564999999999, 42.377017
], "pop" : 36963, "state" : "MA", "_id" : "01002" }
{ "city" : "BARRE", "loc" : [ -72.10835400000001, 42.409698 ],
"pop" : 4546, "state" : "MA", "_id" : "01005" }
{ "city" : "BELCHERTOWN", "loc" : [ -72.41095300000001,
42.275103 ], "pop" : 10579, "state" : "MA", "_id" : "01007" }
{ "city" : "BLANDFORD", "loc" : [ -72.936114, 42.182949 ],
"pop" : 1240, "state" : "MA", "_id" : "01008" }
{ "city" : "BRIMFIELD", "loc" : [ -72.188455, 42.116543 ],
"pop" : 3706, "state" : "MA", "_id" : "01010" }
{ "city" : "CHESTER", "loc" : [ -72.988761, 42.279421 ],
"pop" : 1688, "state" : "MA", "_id" : "01011" }
{ "city" : "CHESTERFIELD", "loc" : [ -72.833309, 42.38167 ],
"pop" : 177, "state" : "MA", "_id" : "01012" }
{ "city" : "CHICOPEE", "loc" : [ -72.607962, 42.162046 ],
"pop" : 23396, "state" : "MA", "_id" : "01013" }
{ "city" : "CHICOPEE", "loc" : [ -72.576142, 42.176443 ],
"pop" : 31495, "state" : "MA", "_id" : "01020" }
{ "city" : "WESTOVER AFB", "loc" : [ -72.558657, 42.196672 ],
"pop" : 1764, "state" : "MA", "_id" : "01022" }
{ "city" : "CUMMINGTON", "loc" : [ -72.905767, 42.435296 ],
"pop" : 1484, "state" : "MA", "_id" : "01026" }
```

In Chapter 8, we learn how to use pandas to create and export data frames in JSON.

Other Formats

We've now seen how to import files and data using some of the most common formats in Python. Other formats include the following:

- lxml—particularly the **objectify** module—allows you to import files into XML.

- SQLite3 allows you to import SQL databases.

- PyMongo allows you to manage Mongo databases

- feedparser allows you to process feeds in many formats, including RSS

- xlrd allows you to import files into Excel (note, however, that pandas is much easier to use)

Summary

Importing a file and a dataset in multiple formats is one of the most important things in data analysis. The procedures described in this chapter are important because we don't need a library to import files and data. We can use the basic functions in Python.

CHAPTER 8

pandas

In Chapter 7, we learned how to import a generic file using basic functions. Here we explore pandas—one of the most important libraries for dataset management.

Libraries for Data Mining

From this point onward, including the following chapters, we examine the most important and most used data mining libraries:

- **pandas:** imports, manages, and manipulates data frames in various formats, extracts part of the data, combines two data frames, and also contains some basic statistical functions

- **NumPy:** a package for scientific computing, contains several high-level mathematical and algebraic functions, and random number generation; and allows the creation of arrays

- **Matplotlib:** a library that allows the creation charts from datasets

- **SciPy:** contains more than 60 statistical functions related to mathematical and statistical analysis

- **scikit-learn:** the most important tool for machine learning and data analysis.

© Valentina Porcu 2018
V. Porcu, *Python for Data Mining Quick Syntax Reference*,
https://doi.org/10.1007/978-1-4842-4113-4_8

pandas

In this chapter, we move away from discussions of Python structures and start to look at the most important data mining packages, beginning with the pandas library.

pandas is an open-source Python library that contains various tools for importing, managing, and manipulating data. It has a number of high-level features for manipulating, reorganizing and scanning structured data, including slicing, managing missing values, restructuring data, extracting dataset parts, and importing and data parsing from the Web. pandas is one of the most important data mining libraries. We can

- Read and import structured data

- Organize and manipulate them

- Calculate some basic statistics

We can import the whole library:

```
>>> import pandas as pd
```

or import only primary structures in pandas, which are Series and DataFrame:

```
>>> from pandas import Series, DataFrame
```

As we shall see, pandas, NumPy, SciPy, and Matplotlib typically work together. I present them in the best possible way—separately at first—to effect clarity.

pandas: Series

As a first action, we always download and install the package or packages we need—in this case, pandas. pandas is part of the Anaconda suite, so you do not have to install it if you have Anaconda installed. Call it this way:

```
>>> import pandas as pd
```

As you can see, we import the package by creating a shortcut: pd. To call a pandas function at this point, we simply write

pd.function_name()

We can also import specific structures, such as Series and DataFrames:

```
>>> from pandas import Series, DataFrame
```

In this case, when we call up or create one of these two structures, we need not specify "pd" at the beginning.

The first pandas structure is Series—a one-dimensional array characterized by an index. Let's create our first series:

```
>>> series1 = Series([25, 27, 28, 30],
        index = ["student1", "student2", "student3", "student4"])

>>> print(series1)

student1    25
student2    27
student3    28
student4    30
dtype: int64

# we then use the .describe() function to get some statistical
information about the series

>>> series1.describe()

count     4.000000
mean     27.500000
std       2.081666
min      25.000000
25%      26.500000
50%      27.500000
```

```
75%        28.500000
max        30.000000
dtype: float64
```

Let's look at the first element of the series:

```
>>> series1[0]
```

```
25
```

We can also extract multiple elements:

```
>>> series1[[2,3]]
```

```
student3    28
student4    30
dtype: int64
```

Or elements from the index:

```
>>> series1["student1"]
25
```

```
>>> series1[["student1", "student4"]]
```

```
student1    25
student4    30
dtype: int64
```

The .index method allows us to index the series:

```
>>> series1.index
```

```
Index(['student1', 'student2', 'student3', 'student4'],
dtype='object')
```

We can also verify the presence of an element in a series:

```
>>> 'student2' in series1
```

```
True
```

124

If we did not import series and data frames separately, we can create a series by specifying that we are using the pandas library:

```
>>> series2 = pd.Series([40, 20, 35, 70],
        index = ["price1", "price2", "price3", "price4"])
```

```
# if we did not import pandas as pd, we can specify the full
package name
```

```
>>> pandas.Series([40, 20, 35, 70],
        index = ["price1", "price2", "price3", "price4"])
```

The pandas series are marked by indexes. If we do not specify the index, it is created automatically:

```
>>> series3 = Series([25, 27, 28, 30])
```

```
series3
0    25
1    27
2    28
3    30
dtype: int64
```

To replace some of the elements in a series, we use the .replace method.

```
>>> series3.replace([25, 27], [125, 127])
```

```
0    125
1    127
2     28
3     30
dtype: int64
```

Let's create another series: one with random numbers. To do this, we need to load the NumPy library:

```
>>> import numpy as np

# we create a series of random numbers

>>> series4 = pd.Series(np.random.randn(7))

# we print it and examine the index

series4
0    -0.227393
1    -1.079208
2     0.101591
3     0.157502
4     1.541307
5    -0.182501
6    -0.247327
dtype: float64

# we can modify the index with the .index() method

>>> series4.index = ['a', 'b', 'c', 'd', 'e', 'f', 'g']

# and check the index again

series4
a    -0.227393
b    -1.079208
c     0.101591
d     0.157502
e     1.541307
f    -0.182501
g    -0.247327
dtype: float64
```

```
# we create another random series

>>> series5 = pd.Series(np.random.randn(7))

# and then merge the two series using the .concat() method

>>> s45 = pd.concat([series4, series5])

# we print the created series

>>> s45

a    -0.227393
b    -1.079208
c     0.101591
d     0.157502
e     1.541307
f    -0.182501
g    -0.247327
0    -0.610507
1     0.282318
2    -1.142692
3    -1.081449
4    -1.818420
5    -1.133354
6     0.804213
dtype: float64

# we check one of the created structures

>>> type(s45)

>>> pandas.core.series.Series
```

```
# we remove some elements of the series

>>> >>> del(s45[5])

>>> s45
a    -0.227393
b    -1.079208
c     0.101591
d     0.157502
e     1.541307
f    -0.182501
g    -0.247327
0    -0.610507
1     0.282318
2    -1.142692
3    -1.081449
4    -1.818420
6     0.804213
dtype: float64

>>> del(s45['f'])

>>> s45
b    -1.079208
c     0.101591
e     1.541307
g    -0.247327
0    -0.610507
1     0.282318
2    -1.142692
3    -1.081449
4    -1.818420
6     0.804213
dtype: float64
```

```
# we slice some elements from the series

>>> s45[0]

-0.61050680805211532

>>> s45[2:4]

e    1.541307
g    -0.247327
dtype: float64

# we can perform object slicing up to the fourth position

>>> s45[:4]

# or from the fifth position onward

>>> s45[5:]

# or we can extract the last three items

>>> s45[-3:]

# we can extract the first and last elements using with the
.head() and .tail() methods

>>> s45.head(2)

>>> s45.tail(3)

# we can reverse the series

>>> s45[::-1]

# last, we can create a copy of the series with the .copy() method

>>> copys45 = s45.copy()
```

pandas: Data Frames

The most important structure in pandas is the DataFrame, a structure which extends the capabilities of the Series and allows us to manage our datasets. Let's create a small data frame with pandas:

```
# we import pandas

>>> import pandas as pd

# we use the DataFrame functions to create or import a dataset
or file from the computer

# to get help with a particular function, we type

>>> help(pd.DataFrame())

Help on DataFrame in module pandas.core.frame object:

# class DataFrame(pandas.core.generic.NDFrame)
 |  Two-dimensional size-mutable, potentially heterogeneous
 |  tabular data structure with labeled axes (rows and columns).
 |  Arithmetic operations align on both row and column labels.
 |  Can be thought of as a dict-like container for Series objects.
 |  The primary pandas data structure
 |
 |  Parameters
 |  ----------
 |  data : numpy ndarray (structured or homogeneous), dict, or
 |  DataFrame
 |      Dict can contain Series, arrays, constants, or list-like
 |      objects
```

```
|   index : Index or array-like
|       Index to use for resulting frame. Will default to
|       np.arange(n) if no indexing information part of input
|       data and no index provided
|   columns : Index or array-like
|       Column labels to use for resulting frame. Will default to
|       np.arange(n) if no column labels are provided
|   dtype : dtype, default None
```

we create our first dataset

```
>>> df1 = pd.DataFrame({'Names': ['Simon', 'Kate', 'Francis',
'Laura', 'Mary', 'Julian', 'Rosie'],
        'Height':[180, 165, 170, 164, 163, 175, 166],
        'Weight':[85, 65, 68, 45, 43, 72, 46],
        'Pref_food':['steak', 'pizza', 'pasta', 'pizza', 'veget
        ables','steak','seafood'],
        'Sex':['m', 'f','m','f', 'f', 'm', 'f']})
```

caution: we cannot create variables with names that contain
a space; for instance, we can create the variable 'Var_1', but
naming a variable 'Var 1' returns an error

we print the data frame

```
>>> df1
   Height    Names  Pref_food Sex  Weight
0     180    Simon      steak   m      85
1     165     Kate      pizza   f      65
2     170  Francis      pasta   m      68
3     164    Laura      pizza   f      45
4     163     Mary vegetables   f      43
5     175   Julian      steak   m      72
6     166    Rosie    seafood   f      46
```

```
# it may be useful to conduct an analysis of our data, which we
can do using the .describe() method
```

```
>>> df1.describe()
```

	Height	Weight
count	7.000000	7.000000
mean	169.000000	60.571429
std	6.377042	16.154021
min	163.000000	43.000000
25%	164.500000	45.500000
50%	166.000000	65.000000
75%	172.500000	70.000000
max	180.000000	85.000000

```
# we can recall variable names
```

```
>>> df1.columns
```

```
Index(['Height', 'Names', 'Pref_food', 'Sex', 'Weight'],
dtype='object')
```

At this point, we can continue by setting a variable—in this case, 'Names'—as index in this way:

```
>>> df1.set_index('Names')
```

Names	Height	Pref_food	Sex	Weight
Simon	180	steak	m	85
Kate	165	pizza	f	65
Francis	170	pasta	m	68
Laura	164	pizza	f	45
Mary	163	vegetables	f	43
Julian	175	steak	m	72
Rosie	166	seafood	f	46

```
# to consolidate the index, we use the argument inplace = True

>>> df1.set_index('Names', inplace = True)

# we can reset the index as follows:

>>> df1.reset_index()
```

We can continue to explore our data.

```
# we check the number of cases and variables

>>> print(df1.shape)
(7, 4)

# and determine variable type

>>> print(df1.dtypes)

Height        int64
Pref_food     object
Sex           object
Weight        int64
dtype: object
```

Here, too, we can see the first and the last elements using the head() and tail() functions. We may want to specify the number of cases we want to display. In this case, we would use parentheses:

```
>>> df1.head()

>>> df1.tail(3)
```

We can also get information about variables using the .info() method:

```
>>> df1.info()

<class 'pandas.core.frame.DataFrame'>
RangeIndex: 7 entries, 0 to 6
Data columns (total 5 columns):
```

```
Height          7 non-null int64
Names           7 non-null object
Pref_food       7 non-null object
Sex             7 non-null object
Weight          7 non-null int64
dtypes: int64(2), object(3)
memory usage: 360.0+ bytes
```

We can then rearrange the data according to one of the variables:

```
>>> df1.sort_values(by = 'Weight')
```

Names	Height	Pref_food	Sex	Weight
Mary	163	vegetables	f	43
Laura	164	pizza	f	45
Rosie	166	seafood	f	46
Kate	165	pizza	f	65
Francis	170	pasta	m	68
Julian	175	steak	m	72
Simon	180	steak	m	85

```
# we rearranged our data based on weight, from the lowest to
the highest value. To reverse these values, from highest to
lowest, we specify the ascending argument as False
```

```
>>> df1.sort_values(by = 'Weight', ascending = False)
```

Names	Height	Pref_food	Sex	Weight
Simon	180	steak	m	85
Julian	175	steak	m	72
Francis	170	pasta	m	68
Kate	165	pizza	f	65
Rosie	166	seafood	f	46
Laura	164	pizza	f	45
Mary	163	vegetables	f	43

```
# if there are any missing values in our dataset (in particular,
in the variable used as the index), we can put them all at the
beginning or end of our display in the following ways
```

```
>>> df1.sort_values(by = 'Weight', na_position = "last")
```

```
>>> df1.sort_values(by = 'Weight', na_position = "first")
```

```
# to rearrange the dataset by index, we can use sort_index,
instead of sort_values, to sort the data in an ascending or
descending way
```

```
>>> df1.sort_index()
```

```
>>> df1.sort_index(ascending= False)
```

	Names	Height	Pref_food	Sex	Weight
6	Rosie	166	seafood	f	46
5	Julian	175	steak	m	72
4	Mary	163	vegetables	f	43
3	Laura	164	pizza	f	45
2	Francis	170	pasta	m	68
1	Kate	165	pizza	f	65
0	Simon	180	steak	m	85

Let's look at some examples of slicing items from a data frame:

```
# from the first column (Python starts counting from 0, so we
use the 0 column), let's slice the first three cases:
```

```
>>> df1[:3]
```

	Names	Height	Pref_food	Sex	Weight
0	Simon	180	steak	m	85
1	Kate	165	pizza	f	65
2	Francis	170	pasta	m	68

```
# or slice from the fourth case to the end
```

```
>>> df1[4:]
    Names   Heigth   Pref_food Sex  Weight
4    Mary    163   vegetables   f     43
5  Julian    175        steak   m     72
6   Rosie    166      seafood    f     46
```

```
# or slice one of the columns
```

```
>>> df1['Names']
```

```
0     Simon
1      Kate
2   Francis
3     Laura
4      Mary
5    Julian
6     Rosie
Name: Names, dtype: object
```

```
# by using single square brackets, as we just did, we can
extract a variable like a Series
```

```
# here's another way to do this
```

```
>>> df1.Names
```

```
0     Simon
1      Kate
2   Francis
3     Laura
4      Mary
5    Julian
6     Rosie
Name: Names, dtype: object
```

```
# a third way is to use double square brackets to  we select a
data frame
>>> df1[['Names']]
```

	Names
0	Simon
1	Kate
2	Francis
3	Laura
4	Mary
5	Julian
6	Rosie

```
# we can also select more columns
>>> df1[['Names', 'Sex']]
```

We can select the value of a variable for a particular element—for example, the third value:

```
>>> df1['Sex'][2]
'm'
```

```
# or some values of some variables
>>> df1['Names'][1:4]
```

```
1       Kate
2     Francis
3       Laura
Name: Names, dtype: object
```

To select items from a data frame we can also use the .loc, .iloc, and .ix methods:

- **.loc:** works through the index

- **.iloc:** extracts via position

- **.ix:** takes both into account

```
# .loc
>>> df1.loc[1]
Names          Kate
Height          165
Pref_food     pizza
Sex               f
Weight           65
Name: 1, dtype: object
# we can extract the elements from the first to the fourth
>>> df1.loc[:3]

# .iloc

# we can extract the element found in the first case and in the
third variable
>>> df1.iloc[0,2]
'steak'
# an alternative to .iloc is the .iat method
>>> df1.iat[1,3]
'f'
# the .at and .iat methods are based on the index
>>> df1.at[2,'Sex']
'm'
```

to the left of the comma, we indicate the cases to be extracted; to the right, the columns or variables (table not shown). To extract all elements, we use a colon

what if we wanted to extract every other case?

```
>>> df1[::2]
```

	Names	Height	Pref_food	Sex	Weight
0	Simon	180	steak	m	85
2	Francis	170	pasta	m	68
4	Mary	163	vegetables	f	43
6	Rosie	166	seafood	f	46

Through Boolean operators we can specify extracting conditions. For instance, we can extract the cases of the dataset in which the height of the subjects (Height) is greater than 170:

```
>>> df1[df1.Height > 170]
     Names  Height  Pref_food  Sex  Weight
0    Simon  180          steak    m      85
5    Julian 175          steak    m      72
```

We can also specify multiple conditions, such as extracting all females who weigh more than 163:

```
>>> df1[(df1.Height > 163) & (df1.Sex == 'f')]
```

Alternatively, we can also use the .query() method:

```
>>> df1.query("Sex != 'f'")
     Names  Height  Pref_food  Sex  Weight
0    Simon    180       steak    m      85
2    Francis  170       pasta    m      68
5    Julian   175       steak    m      72
```

Now let's look at how to rename a column:

```
>>> df1 = df1.rename(columns = {'Pref_food': 'Food'})

>>> df1.head(2)
    Names  Height    Food  Sex  Weight
0   Simon     180   steak    m      85
1    Kate     165   pizza    f      65
```

We can create a copy of our data frame:

```
>>> df2 = df1.copy()

>>> df2
```

	Names	Height	Pref_food	Sex	Weight
0	Simon	180	steak	m	85
1	Kate	165	pizza	f	65
2	Francis	170	pasta	m	68
3	Laura	164	pizza	f	45
4	Mary	163	vegetables	f	43
5	Julian	175	steak	m	72
6	Rosie	166	seafood	f	46

We can also add a column. For example, let's add a column to the df2 dataset that contains height in centimeters:

```
>>> df2['new_col'] = df2.Height/100

>>> df2
    Names  Heigth           Food  Sex  Weight  new_col
0   Simon     180          steak    m      85     1.80
1    Kate     165          pizza    f      65     1.65
2 Francis     170          pasta    m      68     1.70
3   Laura     164          pizza    f      45     1.64
4    Mary     163     vegetables    f      43     1.63
5  Julian     175          steak    m      72     1.75
6   Rosie     166        seafood    f      46     1.66
```

We can also add a column using the .insert() method. The first element, 5, marks the location where we want to insert the new column (the sixth position), the second is the name of the new column, and the third is the value of the new column:

```
>>> df2.insert(5, column = 'new_col2', value = df2.Height/100)

>>> print(df2.head(2))

   Names  Heigth Pref_food Sex  Weight  new_col2  new_col
0  Simon  180         steak   m      85      1.80     1.80
1  Kate   165         pizza   f      65      1.65     1.65
```

Let's return to the df1 dataset without additions. By using .groupby(), we can also aggregate the dataset around one or more variables.

```
>>> df1.groupby('Sex')
<pandas.core.groupby.DataFrameGroupBy object at 0x10c8b2be0>
```

We can create an object that contains aggregated groups according to the 'Sex' variable:

```
>>> grouped1 = df1.groupby('Sex')

# to view the groups, we use group_name.groups

>>> grouped1.groups

{'f': [1, 3, 4, 6], 'm': [0, 2, 5]}
```

We can visualize the number of cases labeled in one way or another.

```
# to see the cases, proceed as follows

>>> for names, groups in grouped1:
...     print(names)
...     print(groups)
```

```
f
     Names   Height          Food  Sex  Weight
 1    Kate      165         pizza   f       65
 3   Laura      164         pizza   f       45
 4    Mary      163    vegetables   f       43
 6   Rosie      166       seafood   f       46
m
      Names   Height    Food  Sex   Weight
 0    Simon      180   steak    m       85
 2  Francis      170   pasta    m       68
 5   Julian      175   steak    m       72
```

note that we only get data belonging to a group

>>> grouped1.get_group('f')

we can aggregate cases according to two variables

>>> grouped2 = df1.groupby(['Sex', 'Pref_food'])

>>> grouped2.groups

```
{('f', 'pizza'): [1, 3],
 ('f', 'seafood'): [6],
 ('f', 'vegetables'): [4],
 ('m', 'pasta'): [2],
 ('m', 'steak'): [0, 5]}
```

we can also determine how many cases fall into each of the groups

>>> grouped2.size()

```
Sex   Pref_food
f     pizza         2
      seafood       1
      vegetables    1
m     pasta         1
      steak         2
dtype: int64
```

```
# we can obtain some descriptive statistics about the data
>>> grouped2.describe()
```

Sex	Pref_food		Height	Weight
		count	2.000000	2.000000
		mean	164.500000	55.000000
		std	0.707107	14.142136
		min	164.000000	45.000000
	pizza	25%	164.250000	50.000000
		50%	164.500000	55.000000
		75%	164.750000	60.000000
		max	165.000000	65.000000
		count	1.000000	1.000000
		mean	166.000000	46.000000
		std	NaN	NaN
		min	166.000000	46.000000
f	seafood	25%	166.000000	46.000000
		50%	166.000000	46.000000
		75%	166.000000	46.000000
		max	166.000000	46.000000

We can also count the frequency of a variable using the .value_counts()
method.

```
>>> df1['Sex'].value_counts()
f    4
m    3
Name: Sex, dtype: int64
```

Returning to the df1 dataset, we can acquire information about the variables using the .info() method.

```
>>> print(df1)
```

```
>>> df1.info()
```

With the .astype() method, we can change the nature of a variable. For example, we can overwrite a variable from int (a numeric variable) to an object variable, such as a string:

```
>>> df1['Weight'] = df1['Weight'].astype('object')
```

```
>>> df1.info()
<class 'pandas.core.frame.DataFrame'>
RangeIndex: 7 entries, 0 to 6
Data columns (total 5 columns):
Height      7 non-null int64
Names       7 non-null object
Pref_food   7 non-null object
Sex         7 non-null object
Weight      7 non-null object
dtypes: int64(1), object(4)
memory usage: 360.0+ bytes
```

We can delete cases and variables with the .drop() method:

```
>>> df1.drop(0)
```

	Names	Height	Food	Sex	Weight
1	Kate	165	pizza	f	65
2	Francis	170	pasta	m	68
3	Laura	164	pizza	f	45
4	Mary	163	vegetables	f	43
5	Julian	175	steak	m	72
6	Rosie	166	seafood	f	46

```
# if, on the other hand, we set and consolidate the index, we
eliminate the case based on the value of the index

>>> df1.set_index('Names', inplace = True)

>>> df1.drop("Laura")
```

	Height	Pref_food	Sex	Weight
Names				
Simon	180	steak	m	85
Kate	165	pizza	f	65
Francis	170	pasta	m	68
Mary	163	vegetables	f	43
Julian	175	steak	m	72
Rosie	166	seafood	f	46

```
# we can remove more than one element at a time

>>> print(df1.drop(["Mary", "Francis"]))

# or remove an entire column

# to do this, we must specify axis 1 (the default axis 0
indicates rows)

>>> df1.drop("Height", axis = 1)
```

Names	Pref_food	Sex	Weight
Simon	steak	m	85
Kate	pizza	f	65
Francis	pasta	m	68
Laura	pizza	f	45
Mary	vegetables	f	43
Julian	steak	m	72
Rosie	seafood	f	46

To remove multiple columns, include them in a list:

```
>>> df.drop(["column1", "column2"], axis = 1)
# instead of specifying the axis as 1, we can also specify
"columns"
>>> df.drop(["column1", "column2"], axis = "columns")
```

The pandas package also allows us to create cross-tabs with the same function. For instance, in the df1 dataset, we can create a cross-tab that correlates gender with favorite food:

```
>>> pd.crosstab(df1.Sex, df1.Pref_food)
```

Pref_food	pasta	pizza	seafood	steak	vegetables
Sex					
f	0	2	1	0	1
m	1	0	0	2	0

When we add the margins parameter, we also get the row with the totals:

```
>>> pd.crosstab(df1.Sex, df1.Pref_food, margins = True)
```

Pref_food	pasta	pizza	seafood	steak	vegetables	All
Sex						
f	0	2	1	0	1	4
m	1	0	0	2	0	3
All	1	2	1	2	1	7

pandas: Importing and Exporting Data

The pandas library is also very important with regard to importing files from your computer in various formats, including .csv:

```
import pandas as pd

>>> df2 = pd.read_csv("file_address.csv")
```

```
# we can import a file by specifying that the first line does
NOT contain variable names
```

```
>>> df2 = pd.read_csv("file_address.csv", header = None)
```

```
# or by specifying which row contains variable names
```

```
>>> df2 = pd.read_csv("file_address.csv", header = 0)
>>> df2 = pd.read_csv("file_address.csv", header = 1)
```

```
# we can also specify column names
```

```
>>> df2 = pd.read_csv("file_address.csv", names = ['var1',
'var2, 'var3, 'var4])
```

```
# or we can import only a part of the columns
```

```
>>> df2 = pd.read_csv("file_address.csv", usecols = [1,2,3])
```

```
# we can also specify the element separating our data
```

```
>>> df2 = pd.read_csv("file_address.csv", sep = " ")
```

```
# we can now take a look at the first records of the data frame
```

```
>>> df2.head()
```

```
# or the last

>>> df2.tail()

# we use parentheses to specify the number of cases to be
displayed

>>> df2.head(9)
>>> df2.tail(5)

# and we can see a summary of the dataset

>>> df2.describe()

>>> df2.info()

>>> df.dtypes() # it tells us our data type
```

The pd.read_csv() function can also be used to import files from the Web. The UCI machine learning repository (https://archive.ics.uci.edu/ml/index.php) features several data sets used for machine learning.

Let's look at an iris dataset. Figures 8-1 through 8-3 and the following steps help us navigate to its tab and copy the link featuring the dataset.

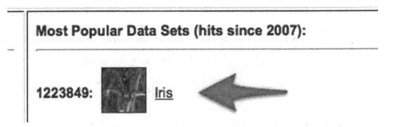

Figure 8-1. *Iris dataset on the UCI machine learning web site click on 'Iris' to get description page*

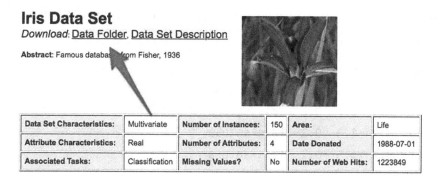

Iris Data Set

Download: Data Folder, Data Set Description

Abstract: Famous database from Fisher, 1936

Data Set Characteristics:	Multivariate	Number of Instances:	150	Area:	Life
Attribute Characteristics:	Real	Number of Attributes:	4	Date Donated	1988-07-01
Associated Tasks:	Classification	Missing Values?	No	Number of Web Hits:	1223849

Figure 8-2. *Iris dataset on the UCI machine learning web site*

Index of /ml/machine-learning-databases/iris

Name	Last modified	Size	Description
Parent Directory		-	
Index	03-Dec-1996 04:01	105	
bezdekIris.data	14-Dec-1999 12:12	4.4K	
iris.data	1993 16:27	4.4K	
iris.names	11-Jul-2000 21:30	2.9K	

Apache/2.2.15 (CentOS) Server at archive.ics.uci.edu Port 443

Figure 8-3. *Iris dataset on the UCI machine learning web site click on 'Data Folder' to access the page containing the dataset*

First, we copy the link featuring the dataset.

```
# we import pandas

>>> import pandas as pd

# we create an object featuring the dataset, which we are
importing with pd.read_csv()

>>> iris = pd.read_csv("https://archive.ics.uci.edu/ml/machine-
learning-databases/iris/iris.data")
```

```
# we check data accuracy by displaying some occurrences using
the function .head()

>>> iris.head(3)
   5.1  3.5  1.4  0.2  Iris-setosa
0  4.9  3.0  1.4  0.2  Iris-setosa
1  4.7  3.2  1.3  0.2  Iris-setosa
2  4.6  3.1  1.5  0.2  Iris-setosa
```

The structure of the pandas function we used, pd.read_csv(), also includes other arguments we did not use for this import:

pd.read_csv(filepath, sep = ',', dtype = None, skiprows = None, index_col = None, skip_blank_lines = True, na_filter = True)

> **filepath** the address of the file (on the computer or externally)
>
> **sep = ','** the separator dividing data, such as a comma or semicolon
>
> **dtype=** a means of specifying column format
>
> **header=** names of variables in the first line, if any
>
> **skiprows=** a means of importing only one part of cases—for example, skiprows = 50 reads data from the 51st case onward
>
> **index_col=** a means of setting a column as a data index
>
> **skip_blank_lines=** a means of removing any blank lines in the dataset
>
> **na_filter=** a means of identifying the missing values in the dataset and, if set to False, removing them

In addition, the **usecols=** argument can be used to import only a few columns in a dataset. For instance, let's say we have the small dataset shown in Figure 8-4, from which we want to extract the following variables:

```
>>> df3 = pd.read_csv("~students.csv",usecols =["ID", "mark1"],
index_col= ["ID"])

# we verify data

df3
     mark1
ID
1      17.0
2      24.0
3      17.0
4      27.0
5      30.0
6      30.0
7      23.0
8      17.0
9      21.0
10     24.0
11     24.0
12     25.0
13     24.0
14      NaN
15     22.0
16     30.0
17     29.0
18     29.0
19      NaN

# correctly, we only loaded the 'mark1' variable
```

```
                            students.csv
"","ID","gender","subject","mark1","mark2","mark3","fres"
"1",1,"M",1,17,20,15,"neg"
"2",2,"F",2,24,30,23,"pos"
"3",3,"F",1,17,16,NA,"neg"
"4",4,"M",3,27,23,21,"pos"
"5",5,"M",2,30,22,24,"pos"
"6",6,"F",1,30,21,25,"pos"
"7",7,"M",2,23,24,24,"pos"
"8",8,"F",3,17,NA,20,"neg"
"9",9,"M",1,21,24,24,"pos"
"10",10,"F",2,24,25,24,"pos"
"11",11,"F",2,24,22,25,"pos"
"12",12,"M",1,25,27,24,"pos"
"13",13,"F",2,24,24,25,"pos"
"14",14,"M",3,NA,17,15,"neg"
"15",15,"M",2,22,27,24,"pos"
"16",16,"F",1,30,24,27,"pos"
"17",17,"F",2,29,27,23,"pos"
"18",18,"M",3,29,26,22,"pos"
"19",19,"F",1,NA,17,15,"neg"
```

Figure 8-4. *Dataset students.csv*

One specular function is df.to_csv("filename"), which allows us to write a .csv file into our work directory. We can specify the **index=False** argument to avoid downloading the index together with data.

As an alternative to this file import function, we can use the pd.read_table(filepath, sep = ',') function, which sets both the file path (filepath) and the separator.

Other pandas features allow us to read Excel .xls or .xlsx files. To do this, we use a generic formula:

df = pd.read_excel(filepath, "sheet_name")

In this case, not only must we specify the address of the file, but also whether the Excel file features more than one data sheet, and the name of the sheet from which we want to read data. As with .csv, for Excel formats we also have a formula that allows us to write an Excel file in the work directory of our computer: df.to_excel(). The pandas package also contains a function for reading files in JSON—pd.read_json()—and also allows us to access Web data via the pd.read_html(url) function and to convert a data frame to an HTML table via pd.to_html().

Let's look at an example of creating and exporting a file in JSON:

```
>>> jf = pd.DataFrame(np.arange(16).reshape(4,4), index = ['A',
'B', 'C', 'D'], columns = ['var1', 'var2', 'var3', 'var4'])

>>> jf
   var1  var2  var3  var4
A     0     1     2     3
B     4     5     6     7
C     8     9    10    11
D    12    13    14    15

>>> jf.to_json('jf.json')
```

```
# all that is left is to check that the file has been created
correctly in our work directory
```

pandas: Data Manipulation

We have seen how to create groups and manipulate data frames. Now let's look at more manipulation elements through pandas. First, let's import pandas and create a small dataset:

```
>>> df2 = pd.DataFrame({'Names': ['Simon', 'Kate', 'Francis',
'Laura', 'Mary', 'Julian', 'Rosie', 'Simon', 'Laura'],
                'Height':[180, 165, 170, 164, 163, 175, 166,
                180, 164],
                'Weight':[85, 65, 68, 45, 43, 72, 46, 85, 45],
                'Pref_food': ['steak', 'pizza', 'pasta',
                'pizza', 'vegetables', 'steak', 'seafood',
                'steak', 'pizza'],
                'Sex': ['m','f','m','f','f','m','f', 'm', 'f']})
```

Starting with the 'Sex' variable, let's create two dummy variables and continue with pandas:

```
>>> df_dummy = pd.get_dummies(df2['Sex'], prefix = 'Sex')
```

```
>>> df_dummy
```

	Sex_f	Sex_m
0	0.0	1.0
1	1.0	0.0
2	0.0	1.0
3	1.0	0.0
4	1.0	0.0
5	0.0	1.0
6	1.0	0.0
7	0.0	1.0
8	1.0	0.0

At this point, we can join these two dummy variables to the original dataset:

```
>>> df2.join(df_dummy)
```

	Unnamed: 0	Height	Names	Food	Sex	Weight	Sex_f	Sex_m
0	0	180	Simon	steak	m	85	0.0	1.0
1	1	165	Kate	pizza	f	65	1.0	0.0
2	2	170	Francis	pasta	m	68	0.0	1.0
3	3	164	Laura	pizza	f	45	1.0	0.0
4	4	163	Mary	vegetables	f	43	1.0	0.0
5	5	175	Julian	steak	m	72	0.0	1.0
6	6	166	Rosie	seafood	f	46	1.0	0.0
7	0	180	Simon	steak	m	85	0.0	1.0
8	3	164	Laura	pizza	f	45	1.0	0.0

We save the joined dataset featuring the dummy variables in a new object:

```
>>> df3 = df2.join(df_dummy)

# let's check

>>> print(df3.head(2))
```

	Height	Names	Pref_food	Sex	Weight	Sex_f	Sex_m
0	180	Simon	steak	m	85	0.0	1.0
1	165	Kate	pizza	f	65	1.0	0.0

Now let's remove the original 'Sex' variable:

```
>>> del df3['Sex']

# double-check

>>> print(df3.head(2))
```

	Height	Names	Pref_food	Weight	Sex_f	Sex_m
0	180	Simon	steak	85	0.0	1.0
1	165	Kate	pizza	65	1.0	0.0

As you can see, there are a couple of duplicate cases inside our dataset. First, we need to identify them:

```
>>> df2.duplicated()
0    False
1    False
2    False
3    False
4    False
5    False
6    False
```

```
7       True
8       True
dtype: bool
```

```
# our dataset has two duplicate cases
```

To remove duplicate cases, we use .drop_duplicates():

```
>>> df2.drop_duplicates()
```

To get a dataset without duplicates, we overwrite the old one or create another object:

```
>>> df3 = df2.drop_duplicates()
```

We can also delete a case, this time using drop:

```
>>> df2.drop(2)
```

	Height	Names	Pref_food	Sex	Weight
0	180	Simon	steak	m	85
1	165	Kate	pizza	f	65
3	164	Laura	pizza	f	45
4	163	Mary	vegetables	f	43
5	175	Julian	steak	m	72
6	166	Rosie	seafood	f	46
7	180	Simon	steak	m	85
8	164	Laura	pizza	f	45

```
# in this case, we delete the third case
```

The stack() and unstack() functions allow us to reorganize our data in a different way:

```
>>> df3.stack()
```

```
0   Height                    180
    Names                   Simon
    Pref_food               steak
    Sex                         m
    Weight                     85
1   Height                    165
    Names                    Kate
    Pref_food               pizza
    Sex                         f
    Weight                     65
2   Height                    170
    Names                 Francis
    Pref_food               pasta
    Sex                         m
    Weight                     68
3   Height                    164
    Names                   Laura
    Pref_food               pizza
    Sex                         f
    Weight                     45
4   Height                    163
    Names                    Mary
    Pref_food          vegetables
    Sex                         f
    Weight                     43
[...]
```

```
# we save the result in an object

>>> stacked = df3.stack()
```

```
# and now we can return it to its original format using
unstacked()
```

```
>>> unstacked = stacked.unstack()
```

```
>>> print(unstacked)
```

	Height	Names	Pref_food	Sex	Weight
0	180	Simon	steak	m	85
1	165	Kate	pizza	f	65
2	170	Francis	pasta	m	68
3	164	Laura	pizza	f	45
4	163	Mary	vegetables	f	43
5	175	Julian	steak	m	72
6	166	Rosie	seafood	f	46

We can also reorganize our data via the melt() function:

```
>>> pd.melt(df3)
```

	variable	value
0	Height	180
1	Height	165
2	Height	170
3	Height	164
4	Height	163
5	Height	175
6	Height	166
7	Names	Simon
8	Names	Kate
9	Names	Francis

```
10       Names       Laura
11       Names        Mary
12       Names      Julian
13       Names       Rosie
```

[...]

Let's go back to the initial dataset

```
>>> df2.head(2)
   Height  Names  Pref_food Sex  Weight
0     180  Simon      steak   m      85
1     165   Kate      pizza   f      65
```

and use the .T() function to transpose rows and columns:

```
# we invert rows and columns
```

```
>>> df2.T
```

	0	1	2	3	4	5	6	7	8
Height	180	165	170	164	163	175	166	180	164
Names	Simon	Kate	Francis	Laura	Mary	Julian	Rosie	Simon	Laura
Pref_food	steak	pizza	pasta	pizza	vegetables	steak	seafood	steak	pizza
Sex	m	f	m	f	f	m	f	m	f
Weight	85	65	68	45	43	72	46	85	45

We can also extract a random sample of our data using the .sample() function:

```
>>> df2.sample(n=2)

   Height  Names   Pref_food Sex  Weight
6     166  Rosie     seafood   f      46
4     163   Mary  vegetables   f      43
```

In parentheses (n=2), we inserted the number of cases to be extracted. How can we always extract the same cases randomly, so that extraction can be repeated? We use the np.random.seed() function. The number included in parentheses in this function does not really matter, but if two people use the same dataset and use the same number, they will extract the same cases.

To do this, we need to import the NumPy package:

```
>>> import numpy as np

>>> np.random.seed(1)

>>> df2.sample(n=2)
```

	Height	Names	Pref_food	Sex	Weight
3	164	Laura	pizza	f	45
7	180	Simon	steak	m	85

```
>>> np.random.seed(1)
>>> df2.sample(n=2)
```

	Height	Names	Pref_food	Sex	Weight
3	164	Laura	pizza	f	45
7	180	Simon	steak	m	85

Instead of extracting a number of cases, we can extract a percentage, with the argument **frac=** instead of **n=**

```
>>> df2.sample(frac = .1)

>>> df2.sample(frac = .5)
```

Using the argument **frac =**, we specified the percentage; 0.1 is 10%, 0.5 is 50%, and so on.

```
# we can view cases with the highest values for a certain
variable by specifying the number of cases (in this case, 3)
and the column (in this case, Weight)
```

```
# let's consider the df3 dataset, which does not feature duplicates

>>> df3.nlargest(3, "Weight")
   Height    Names Pref_food Sex  Weight
0     180    Simon     steak   m      85
5     175   Julian     steak   m      72
2     170  Francis     pasta   m      68
```

Let's now call the cases with the lowest values for a certain variable, and specify the number of cases:

```
>>> df3.nsmallest(4, "Weight")

   Height  Names   Pref_food Sex  Weight
4     163   Mary  vegetables   f      43
3     164  Laura       pizza   f      45
6     166  Rosie     seafood   f      46
1     165   Kate       pizza   f      65
```

Last, we can create a small dataset and reorganize data with the pivot_table() function:

```
>>> df4 = pd.DataFrame({'Date': ['2017-01-01', '2017-01-01',
'2017-01-02', '2017-01-01', '2017-01-02', '2017-01-02', '2017-
01-03', '2017-01-02', '2017-01-03', '2017-01-03'],
                        'Type':['x', 'x', 'y', 'x', 'y', 'x', 'z',
                        'y', 'z', 'y'],
                        'Value':[185, 265, 168, 245, 143, 172, 346,
                        285, 145, 128],
                        })

>>> print(df4)
```

```
        Date Type  Value
0  2017-01-01    x    185
1  2017-01-01    x    265
2  2017-01-02    y    168
3  2017-01-01    x    245
4  2017-01-02    y    143
5  2017-01-02    x    172
6  2017-01-03    z    346
7  2017-01-02    y    285
8  2017-01-03    z    145
9  2017-01-03    y    128
```

```
>>> print(pd.pivot_table(df4, index = "Date", values = "Value",
columns = "Type"))
```

```
Type                  x              y       z
Date
2017-01-01   231.666667           NaN     NaN
2017-01-02   172.000000    198.666667     NaN
2017-01-03          NaN    128.000000   245.5
```

```
# we aggregate the values around the dates
```

pandas: Missing Values

Now let's examine how we can manage datasets with missing data. After importing pandas, we load a dataset that has missing values:

```
>>> import pandas as pd
>>> df_missing = pd.read_csv('df_missing.csv')
```

```
# I created this tiny dataset with missing data, as you can see
```

```
>>> df_missing
```

```
      A     B      C       D    E
0    15  96.0   74.0    31.0   50
1    41  27.0   74.0   279.0   57
2    21  32.0    NaN    99.0   96
3    48  97.0   50.0     NaN   69
4    63  98.0   74.0    44.0   55
5    43  11.0   80.0    74.0   33
6    39  38.0   81.0    20.0   41
7    58  31.0    NaN    76.0   91
8    85  94.0   37.0    65.0   60
9    98   NaN   19.0    43.0   32
```

We can verify the number of cases and variables:

```
>>> print(df_missing.shape)
(10, 5)
```

We can check the complete cases:

```
>>> print(df_missing.dropna().shape)
(6, 5)
```

We can also use .isnull(), which answers the question: Is a value missing?

```
>>> pd.isnull(df_missing)
       A      B      C      D      E
0  False  False  False  False  False
1  False  False  False  False  False
2  False  False   True  False  False
3  False  False  False   True  False
4  False  False  False  False  False
5  False  False  False  False  False
6  False  False  False  False  False
7  False  False   True  False  False
8  False  False  False  False  False
9  False   True  False  False  False
```

The converse, .notnull(), answers the question: Is a value *not* missing?

```
>>> pd.notnull(df_missing)
```

	A	B	C	D	E
0	True	True	True	True	True
1	True	True	True	True	True
2	True	True	False	True	True
3	True	True	True	False	True
4	True	True	True	True	True
5	True	True	True	True	True
6	True	True	True	True	True
7	True	True	False	True	True
8	True	True	True	True	True
9	True	False	True	True	True

We can add the missing values per column:

```
>>> df_missing.isnull().sum()
A    0
B    1
C    2
D    1
E    0
```

```
# or determine the total of missing data
```

```
>>> df_missing.isnull().sum().sum()
4
```

Missing values can be treated by deletion (or by deleting the cases that contain them) or imputation (that is, by replacing missing values with other values, such as a fixed value or an average). To perform a deletion, proceed as follows:

```
>>> df_missing.dropna()
    A     B     C      D   E
0  15  96.0  74.0   31.0  50
1  41  27.0  74.0  279.0  57
4  63  98.0  74.0   44.0  55
5  43  11.0  80.0   74.0  33
6  39  38.0  81.0   20.0  41
8  85  94.0  37.0   65.0  60
```

We eliminated the lines containing missing values. We can also delete the columns with missing values:

```
>>> df_missing.dropna(axis = 1, how = 'any')
    A   E
0  15  50
1  41  57
2  21  96
3  48  69
4  63  55
5  43  33
6  39  41
7  58  91
8  85  60
9  98  32
```

As mentioned, using imputation methods, we can replace a missing value with another value. Let's replace all missing values with a fixed value—in this case, zero:

```
>>> df_missing.fillna(0)
     A    B     C      D    E
0   15  96.0  74.0   31.0  50
1   41  27.0  74.0  279.0  57
2   21  32.0   0.0   99.0  96
3   48  97.0  50.0    0.0  69
4   63  98.0  74.0   44.0  55
5   43  11.0  80.0   74.0  33
6   39  38.0  81.0   20.0  41
7   58  31.0   0.0   76.0  91
8   85  94.0  37.0   65.0  60
9   98   0.0  19.0   43.0  32
```

We can also replace missing values with a word, such as "missing":

```
>>> df_missing.fillna('missing')
     A        B        C        D   E
0   15       96       74       31  50
1   41       27       74      279  57
2   21       32  missing       99  96
3   48       97       50  missing  69
4   63       98       74       44  55
5   43       11       80       74  33
6   39       38       81       20  41
7   58       31  missing       76  91
8   85       94       37       65  60
9   98  missing       19       43  32
```

Or impute missing values to the average:

```
>>> df_missing.fillna(df_missing.mean())
     A          B       C          D   E
0   15  96.000000  74.000   31.000000  50
1   41  27.000000  74.000  279.000000  57
2   21  32.000000  61.125   99.000000  96
3   48  97.000000  50.000   81.222222  69
4   63  98.000000  74.000   44.000000  55
5   43  11.000000  80.000   74.000000  33
6   39  38.000000  81.000   20.000000  41
7   58  31.000000  61.125   76.000000  91
8   85  94.000000  37.000   65.000000  60
9   98  58.222222  19.000   43.000000  32
```

Let's consider a variable, such as variable 'C'. The following line of code will replace every missing value in the variable C with the mean of C:

```
>>> df_missing['C'].fillna(df_missing['C'].mean())
0    74.000
1    74.000
2    61.125
3    50.000
4    74.000
5    80.000
6    81.000
7    61.125
8    37.000
9    19.000
Name: C, dtype: float64
```

Two important .fillna methods can be used to impute missing values: ffill and backfill:

```
>>> df_missing['C'].fillna(method = 'ffill')
0    74.0
1    74.0
2    74.0
3    50.0
4    74.0
5    80.0
6    81.0
7    81.0
8    37.0
9    19.0
Name: C, dtype: float64
>>> df_missing['C'].fillna(method = 'backfill')
0    74.0
1    74.0
2    50.0
3    50.0
4    74.0
5    80.0
6    81.0
7    37.0
8    37.0
9    19.0
Name: C, dtype: float64

# ffill replaces the missing value with the default value
preceding it, whereas backfill replaces it with the nonmissing
value that follows it. To save the results of one of these
insertion or replacement operations, we must create a new data
frame or overwrite the initial one.
```

pandas: Merging Two Datasets

There are many ways to combine two data frames. We can delete overlapping cases, juxtapose two files, and join new cases or new variables. Let's look at some pandas examples. First, we need to create sample data frames:

```
>>> df1 = pd.DataFrame({'id': ['A', 'B', 'C', 'D'], 'var1' :
[37, 36, 43, 23], 'var2': [120, 117, 230, 315]})
```

```
>>> df1
  id  var1  var2
0  A    37   120
1  B    36   117
2  C    43   230
3  D    23   315
```

```
>>> df2 = pd.DataFrame({'id': ['A', 'B', 'C', 'D'], 'var3' :
[12, 16, 13, 18], 'var4': [75, 54, 21, 36]})
```

```
>>> df2
  id  var3  var4
0  A    12    75
1  B    16    54
2  C    13    21
3  D    18    36
```

```
>>> df3 = pd.DataFrame({'id': ['A', 'A', 'B', 'B'], 'var3' :
[2, 6, 3, 8], 'var4': [7, 5, 2, 6]})
```

```
  id  var3  var4
0  A     2     7
1  A     6     5
2  B     3     2
3  B     8     6
```

The easiest way to merge two datasets is to use the .append() method:

```
>>> df1.append(df2)
```

	id	var1	var2	var3	var4
0	A	37.0	120.0	NaN	NaN
1	B	36.0	117.0	NaN	NaN
2	C	43.0	230.0	NaN	NaN
3	D	23.0	315.0	NaN	NaN
0	A	NaN	NaN	12.0	75.0
1	B	NaN	NaN	16.0	54.0
2	C	NaN	NaN	13.0	21.0
3	D	NaN	NaN	18.0	36.0

```
>>> df3.append(df1)
```

	id	var1	var2	var3	var4
0	A	NaN	NaN	2.0	7.0
1	A	NaN	NaN	6.0	5.0
2	B	NaN	NaN	3.0	2.0
3	B	NaN	NaN	8.0	6.0
0	A	37.0	120.0	NaN	NaN
1	B	36.0	117.0	NaN	NaN
2	C	43.0	230.0	NaN	NaN
3	D	23.0	315.0	NaN	NaN

As you can see, the two datasets are juxtaposed every time, without the system taking into account and adjusting the index . To merge more than two datasets, we can use .concat():

```
>>> pd.concat([df1, df2, df3])
```

	id	var1	var2	var3	var4
0	A	37.0	120.0	NaN	NaN
1	B	36.0	117.0	NaN	NaN
2	C	43.0	230.0	NaN	NaN
3	D	23.0	315.0	NaN	NaN
0	A	NaN	NaN	12.0	75.0
1	B	NaN	NaN	16.0	54.0
2	C	NaN	NaN	13.0	21.0
3	D	NaN	NaN	18.0	36.0
0	A	NaN	NaN	2.0	7.0
1	A	NaN	NaN	6.0	5.0
2	B	NaN	NaN	3.0	2.0
3	B	NaN	NaN	8.0	6.0

In this way, we concatenated the datasets horizontally. We can also concatenate them vertically by specifying the axis as 1 (in the previous example, the axis was not specified so it defaulted to zero):

```
>>> pd.concat([df1, df2, df3], axis = 1)
```

```
# the code above and this line below will give us the same result
```

```
>>> pd.concat([df1, df2, df3], axis = 'columns')
```

	id	var1	var2	id	var3	var4	id	var3	var4
0	A	37	120	A	12	75	A	2	7
1	B	36	117	B	16	54	A	6	5
2	C	43	230	C	13	21	B	3	2
3	D	23	315	D	18	36	B	8	6

We can combine two datasets in an even more advanced way, starting from two tables that have some rows and columns in common, and merging them through shared data. The pandas function that allows us to make this type of join is merge(). First, we need to create two datasets:

```
>>> stud1 = pd.DataFrame({'ID': [1,2,3,4,5], 'names' : ['John',
'Greta', 'Francis', 'Laura', 'Charles'], 'Logic': [28, 27, 23,
25, 30]})
```

```
>>> stud2 = pd.DataFrame({'ID': [1,6,3,4,7], 'names' : ['John',
'Kate', 'Francis', 'Laura', 'Simon'], 'Analysis': [23, 24, 28,
29, 22]})
```

```
# by default, the merge() function extracts common cases to the
two datasets, so it is 'inner' by default. Inner join selects
records that have matching values in both datasets
```

```
>>> pd.merge(stud1, stud2, on = 'ID')
```

	ID	Logic	names_x	Analysis	names_y
0	1	28	John	23	John
1	3	23	Francis	28	Francis
2	4	25	Laura	29	Laura

There are other means of merging two datasets: 'left', 'right', and 'outer'. We use the dataset index—in this case, the ID column—as a key (index) to merge the datasets.

```
>>> print(pd.merge(stud1, stud2, on = 'ID', how = 'left'))
```

	ID	Logic	names_x	Analysis	names_y
0	1	28	John	23.0	John
1	2	27	Greta	NaN	NaN
2	3	23	Francis	28.0	Francis
3	4	25	Laura	29.0	Laura
4	5	30	Charles	NaN	NaN

The 'left' parameter allows us to merge the second dataset into the first, taking only the cases of the first dataset and the cases that are common to the second and the first dataset.

```
>>> print(pd.merge(stud1, stud2, on = 'ID', how = 'right'))
```

	ID	Logic	names_x	Analysis	names_y
0	1	28.0	John	23	John
1	3	23.0	Francis	28	Francis
2	4	25.0	Laura	29	Laura
3	6	NaN	NaN	24	Kate
4	7	NaN	NaN	22	Simon

The 'right' parameter allows to merge the first dataset into the second, taking only the cases of the second and those that are common to the second and the first at the same time.

```
>>> print(pd.merge(stud1, stud2, on = 'ID', how = 'outer'))
```

	ID	Logic	names_x	Analysis	names_y
0	1	28.0	John	23.0	John
1	2	27.0	Greta	NaN	NaN
2	3	23.0	Francis	28.0	Francis
3	4	25.0	Laura	29.0	Laura
4	5	30.0	Charles	NaN	NaN
5	6	NaN	NaN	24.0	Kate
6	7	NaN	NaN	22.0	Simon

When we use 'outer', we merge the cases of the two data frames, holding only a copy of the double cases, but without losing data from one of the two initial datasets.

pandas: Basic Statistics

The pandas library has a lot of methods for statistics, which we can use to get descriptive statistical information. As always, we use our data frame:

```
>>> df = pd.DataFrame({'Names': ['Simon', 'Kate', 'Francis',
'Laura', 'Mary', 'Julian', 'Rosie'],
        'Height':[180, 165, 170, 164, 163, 175, 166],
        'Weight':[85, 65, 68, 45, 43, 72, 46],
        'Pref_food':['steak', 'pizza', 'pasta', 'pizza',
        'vegetables','steak','seafood'],
        'Sex':['m', 'f','m','f', 'f', 'm', 'f']})

# some information on numeric data

>>> df.describe()
           Height       Weight
count    7.000000     7.000000
mean   169.000000    60.571429
std      6.377042    16.154021
min    163.000000    43.000000
25%    164.500000    45.500000
50%    166.000000    65.000000
75%    172.500000    70.000000
max    180.000000    85.000000
```

```
# we can also rearrange the statistics with the .transpose
method to make it more readable

>>> df.describe().transpose()
         count        mean        std     min    25%    50%    75%    max
Height     7.0  169.000000   6.377042   163.0  164.5  166.0  172.5  180.0
Weight     7.0   60.571429  16.154021    43.0   45.5   65.0   70.0   85.0
```

```
# value counts

>>> df.count()
Height    7
Names     7
Food      7
Sex       7
Weight    7
dtype: int64

# median calculation

>>> df['Height'].median()
166.0

# average calculation

>>> df['Height'].mean()
169.0

# minimum value

>>> df['Height'].min()
163

# maximum value

>>> df['Height'].max()
180
```

Table 8-1 provides a summary of statistical methods.

Table 8-1. *Statistical Methods*

Method	Description
.describe	Provides some descriptive statistics
.count	Returns the number of values per variable
.mean	Returns the average
.median	Returns the median
.mode	Returns the mode
.min	Returns the lowest value
.max	Returns the highest value
.std	Returns the standard deviation
.var	Returns the variance
.skew	Returns skewness
.kurt	Returns kurtosis

Summary

In this chapter, we learned some easier ways to import and manage our data using pandas—one of the most important libraries for data manipulation and data science. In Chapter 9, we examine another package that is important for data manipulation: NumPy.

CHAPTER 9

SciPy and NumPy

Although pandas is a very important package for data analysis, it can work
in conjunction with two other packages: SciPy and NumPy.

SciPy

SciPy is one of the most important packages for mathematical and statistical
analysis in Python, and it is linked closely to NumPy. SciPy contains more
than 60 statistical functions organized in families of modules:

- scipy.cluster

- scipy.constants

- scipy.fftpack

- scipy.integrate

- scipy.interpolate

- scipy.io

- scipy.lib

- scipy.linalg

- scipy.misc

- scipy.optimize

© Valentina Porcu 2018
V. Porcu, *Python for Data Mining Quick Syntax Reference*,
https://doi.org/10.1007/978-1-4842-4113-4_9

- scipy.signal

- scipy.sparse

- scipy.spatial

- scipy.special

- scipy.stats

- scipy.weave

```
>>> import scipy as sp
```

```
>>> from scipy import stats
```

```
>>> from scipy import cluster
```

We can get help with these modules by typing, for example:

```
>>> help(sp.cluster)
```

```
# or
```

```
>>> help(scipy.cluster)
```

Documentation opens directly in the window, from which we can exit by pressing q.

For instance, SciPy can be used to measure probability density on a number or distribution

```
>>> from scipy.stats import norm
```

```
>>> norm.pdf(5)
1.4867195147342979e-06
```

or an allotment function

```
>>> norm.cdf(x)
```

SciPy features very technical modules. For the purposes of this book, we are particularly interested in combining with NumPy, so the focus of this chapter is on this second package primarily.

NumPy

At the beginning of Python's development, programmers soon found themselves having to incorporate tools for scientific computation. Their first attempt resulting in the Numeric package, which was developed in 1995, followed by an alternative package called Numarray. The merging of the functions of these two packages came to life in 2006 with NumPy.

NumPy stands for "numeric Python" and is an open-source library dedicated to scientific computing, especially with regard to array management. It is sometimes considered as MATLAB version for Python, and features several high-level mathematical functions in algebra and random number generation.

```
>>> import numpy as np
```

```
>>> from numpy import *
```

NumPy's most important structure is a particular type of multidimensional array, called *ndarray*. ndarray consists of two elements: data (the true ndarray) and metadata describing data (dtype or data type). Each ndarray is associated with one and only one dtype. We can have one-dimension arrays:

```
>>> arr1 = np.array([0,1,2,3,4])
```

```
>>> arr1
array([0, 1, 2, 3, 4])
```

Or multidimensional arrays:

```
>>> arr2 = np.array([[5,6,7,8,9], [10,11,12,13,14]])
>>> arr2
array([[ 5,  6,  7,  8,  9],
       [10, 11, 12, 13, 14]])
>>> arr2_1 = np.array([[5,6,7,8,9], [10,11,12,13,14], [2,5,7]])
# the important thing is that data have the same type
```

We can display the data type using the .dtype() function.

```
>>> arr1.dtype
dtype('int64')

# the amount of items in the array

>>> arr1.size
5

# the number of bits

>>> arr1.itemsize
8
```

For instance, by using the NumPy arrange() function, we can create a list of numbers—in the following case, from 0 to 100:

```
>>> arr3 = np.arange(100)
>>> arr3
array([ 0,  1,  2,  3,  4,  5,  6,  7,  8,  9, 10, 11, 12, 13,
14, 15, 16,
       17, 18, 19, 20, 21, 22, 23, 24, 25, 26, 27, 28, 29, 30,
       31, 32, 33,
       34, 35, 36, 37, 38, 39, 40, 41, 42, 43, 44, 45, 46, 47,
       48, 49, 50,
```

```
51, 52, 53, 54, 55, 56, 57, 58, 59, 60, 61, 62, 63, 64,
65, 66, 67,
68, 69, 70, 71, 72, 73, 74, 75, 76, 77, 78, 79, 80, 81,
82, 83, 84,
85, 86, 87, 88, 89, 90, 91, 92, 93, 94, 95, 96, 97, 98, 99])
```

```
# arange also allows us to set a range of numbers
```

```
>>> np.arange(1,10)
array([1, 2, 3, 4, 5, 6, 7, 8, 9])
```

Let's create a fourth array:

```
>>> arr4 = np.array([['a','b','c','d','e'], ['f', 'g', 'h',
'i','l'], ['m', 'n', 'o', 'p','q']])
```

```
>>> arr4
array([['a', 'b', 'c', 'd', 'e'],
       ['f', 'g', 'h', 'i', 'l'],
       ['m', 'n', 'o', 'p', 'q']],
      dtype='<U1')
```

We can select, for example, the first element (0):

```
>>> arr4[0]
```

```
array(['a', 'b', 'c', 'd', 'e'],
      dtype='<U1')
```

If we want to select the third element of the first element, we proceed as follows:

```
>>> arr4[0][3]
'd'
```

Remember that, in Python, we start counting from zero, not one. So, if we select the first element, we actually get the second:

```
>>> arr4[1]
array(['f', 'g', 'h', 'i', 'l'],
      dtype='<U1')

>>> arr4[1][1]
'g'
```

We can also use a negative index and rotate the array as follows:

```
>>> arr4[::-1]
array([['m', 'n', 'o', 'p', 'q'],
       ['f', 'g', 'h', 'i', 'l'],
       ['a', 'b', 'c', 'd', 'e']],
      dtype='|S1')
```

The arr4 array is composed of three different elements. To merge them, we can use the .ravel() method:

```
>>> arr4.ravel()

array(['a', 'b', 'c', 'd', 'e', 'f', 'g', 'h', 'i', 'l', 'm',
'n', 'o','p', 'q'],
      dtype='|S1')
```

Let's create another array, arr5:

```
>>> arr5 = np.array([19, 76, 2, 13, 48, 986, 1, 18])

>>> arr5
array([19, 76, 2, 13, 48, 986, 1, 18])

# we can reorder it from the lowest value to the highest

>>> np.msort(arr5)
array([1, 2, 13, 18, 19, 48, 76, 986])
```

We can reorganize the data of an array using various functions. Let's use the last array created, arr5:

```
# reshape() allows us to reorganize the data of an array–in the
following example, in four cases of two columns each
```

```
>>> arr5.reshape(4,2)
```

```
>>> array([[ 19,  76],
       [  2,  13],
       [ 48, 986],
       [  1,  18]])
```

```
>>> arr5.reshape(8,1)
array([[ 19],
       [ 76],
       [  2],
       [ 13],
       [ 48],
       [986],
       [  1],
       [ 18]])
```

```
# we create two more arrays, organizing them in three columns
from three cases
```

```
>>> x = np.array([20, 42, 17, 3, 7, 12, 14, 70, 9])
```

```
>>> x = x.reshape(3,3)
```

```
>>> x
array([[20, 42, 17],
       [ 3,  7, 12],
       [14, 70,  9]])
```

```
>>> y = x * 3
```

```
>>> y
array([[ 60, 126,  51],
       [  9,  21,  36],
       [ 42, 210,  27]])
```

```
# similar to reshape is the resize() function
```

```
>>> z = np.array([120, 72, 37, 43, 57, 12, 54, 20, 9])
z
array([120,  72,  37,  43,  57,  12,  54,  20,   9])
```

```
>>> z.resize(3,3)
```

```
>>> z
array([[120,  72,  37],
       [ 43,  57,  12],
       [ 54,  20,   9]])
```

```
# we can concatenate two arrays horizontally
```

```
>>> np.hstack((x, y))
array([[ 20,  42,  17,  60, 126,  51],
       [  3,   7,  12,   9,  21,  36],
       [ 14,  70,   9,  42, 210,  27]])
```

```
# we get the same result using the function .concatenate() by
specifying the axis
```

```
>>> np.concatenate((x,y), axis = 1)
array([[ 20,  42,  17,  60, 126,  51],
       [  3,   7,  12,   9,  21,  36],
       [ 14,  70,   9,  42, 210,  27]])
```

```
# or we arrange the data vertically
```

```
>>> np.vstack((x,y))
array([[ 20,  42,  17],
       [  3,   7,  12],
       [ 14,  70,   9],
       [ 60, 126,  51],
       [  9,  21,  36],
       [ 42, 210,  27]])
```

we get the same result with the concatenate() function, without specifying the axis, or by specifying it as zero

```
>>> np.concatenate((x,y))
array([[ 20,  42,  17],
       [  3,   7,  12],
       [ 14,  70,   9],
       [ 60, 126,  51],
       [  9,  21,  36],
       [ 42, 210,  27]])
```

the function .dstack() divides the array into tuples along the third axis

```
>>> np.dstack((x,y))
array([[[ 20,  60],
        [ 42, 126],
        [ 17,  51]],

       [[  3,   9],
        [  7,  21],
        [ 12,  36]],

       [[ 14,  42],
        [ 70, 210],
        [  9,  27]]])
```

```
# the .hsplit() function divides the array into equal parts by
size and shape (in this case, three parts)
```

```
>>> np.hsplit(x, 3)
[array([[20],
       [ 3],
       [14]]), array([[42],
       [ 7],
       [70]]), array([[17],
       [12],
       [ 9]])]
```

```
# we can also use the .split() function
```

```
>>> np.split(x, 3)
[array([[20, 42, 17]]), array([[ 3,  7, 12]]), array([[14,
70,  9]])]
```

```
# or the .vsplit() function, which means vertical splitting
```

```
# we can also convert an array to a list
```

```
>>> z.tolist()
[[120, 72, 37], [43, 57, 12], [54, 20, 9]]
```

As we have seen, NumPy handles both numeric data in various formats and strings, but we can also generate random data. The zeros() and ones() functions create arrays with zeros or ones only.

```
# the first element indicates the number of cases; the second,
the number of variables
```

```
>>> np.zeros((4,3))
array([[ 0.,   0.,   0.],
       [ 0.,   0.,   0.],
       [ 0.,   0.,   0.],
       [ 0.,   0.,   0.]])
```

```
>>> np.ones((5,2))
array([[ 1.,   1.],
       [ 1.,   1.],
       [ 1.,   1.],
       [ 1.,   1.],
       [ 1.,   1.]])
```

NumPy has many ways to store data. Table 9-1 includes some of them from documentation available at https://docs.scipy.org/doc/.

Table 9-1. *Options for Storing Data*

Data Type	Description
bool_	Boolean (True or False), stored as a byte
int_	Default integer type (same as C long; normally either int64 or int32)
intc	Identical to C int (normally int32 or int64)
intp	Integer used for indexing (same as C ssize_t; normally either int32 or int64)
int8	Byte (-128 to 127)
int16	Integer (-32768 to 32767)
int32	Integer (-2147483648 to 2147483647)
int64	Integer (-9223372036854775808 to 9223372036854775807)
uint8	Unsigned integer (0 to 255)
uint16	Unsigned integer (0 to 65535)
uint32	Unsigned integer (0 to 4294967295)
uint64	Unsigned integer (0 to 18446744073709551615)
float_	Shorthand for float64

(*continued*)

Table 9-1. (*continued*)

Data Type	Description
float16	Half-precision float: sign bit, 5-bit exponent, 10-bit mantissa
float32	Single-precision float: sign bit, 8-bit exponent, 23-bit mantissa
float64	Double-precision float: sign bit, 11-bit exponent, 52-bit mantissa
complex_	Shorthand for complex128
complex64	Complex number, represented by two 32-bit floats (real and imaginary components)
complex128	Complex number, represented by two 64-bit floats (real and imaginary components)

When we create an object, we can specify which type of object we want to create:

```
>>> o1 = np.arange(10, dtype = 'int16')
>>> o1
array([0, 1, 2, 3, 4, 5, 6, 7, 8, 9], dtype=int16)
```

NumPy can also be used to create matrices and matrix calculations.

```
# we create two mat1 and mat2 matrices
>>> mat1 = np.matrix([[10, 11, 12, 13, 14], [15, 16, 17, 18, 19], [20, 21, 22, 23, 24]])
>>> mat1
matrix([[10, 11, 12, 13, 14],
        [15, 16, 17, 18, 19],
        [20, 21, 22, 23, 24]])
>>> mat2 = np.matrix([[25, 26, 27, 28, 29], [30, 31, 32, 33, 34], [35, 36, 37, 38, 39]])
```

```
>>> mat2
matrix([[25, 26, 27, 28, 29],
        [30, 31, 32, 33, 34],
        [35, 36, 37, 38, 39]])

# we can do some mathematical operations

>>> mat1 + mat2
matrix([[35, 37, 39, 41, 43],
        [45, 47, 49, 51, 53],
        [55, 57, 59, 61, 63]])

>>> mat2 - mat1
matrix([[15, 15, 15, 15, 15],
        [15, 15, 15, 15, 15],
        [15, 15, 15, 15, 15]])

>>> mat2 / mat1
matrix([[ 2.5       , 2.36363636, 2.25      , 2.15384615, 2.07142857],
        [ 2.        , 1.9375    , 1.88235294, 1.83333333, 1.78947368],
        [ 1.75      , 1.71428571, 1.68181818, 1.65217391, 1.625     ]])

>>> mat1 * 3
matrix([[30, 33, 36, 39, 42],
        [45, 48, 51, 54, 57],
        [60, 63, 66, 69, 72]])

# Numpy lets you calculate the mean

>>> np.mean(mat1)
17.0

# maximum value

>>> np.max(mat1)
24
```

```
# minimum value
```

```
>>> np.min(mat1)
10
```

```
# median
```

```
>>> np.median(mat1)
17.0
```

```
# variance
```

```
>>> np.var(mat1)
18.666666666666668
```

```
# standard deviation
```

```
>>> np.std(mat1)
4.3204937989385739
```

```
# covariance can be calculated by np.cov()
```

NumPy also allows you to import numeric data using the loadtxt() function:

```
>>> up1 = np.loadtxt('~/Python_test/df2', delimiter = ',',
usecols = (1,), unpack = True)
```

```
up1
array([ 0.        ,  15.98293881,  41.7360094 ,  21.02081375,
        54.06254967,   6.68691812,  43.83810058,  39.55058136,
        58.04370289,  85.02891662,  98.25872709])
```

```
# to import more columns, we change the usecols argument:
```

```
>>> up1 = np.loadtxt('~/Python_test/df2', delimiter = ',',
usecols = (1,2,3,4,5), unpack = True)
```

NumPy: Generating Random Numbers and Seeds

To generate random numbers, we need to import the NumPy package.

```
>>> import numpy as np
```

We can generate a random number as follows:

```
>>> np.random.rand()
0.992777076172216
```

```
# or specify in parentheses the number of rows and columns to
be generated automatically

>>> np.random.rand(2,3)
array([[ 0.39352349,  0.57116926,  0.88967038],
       [ 0.76375617,  0.24620255,  0.17408501]])
```

We can create multidimensional arrays of random numbers:

```
# to a size (10 is the upper limit from which the distribution
is extracted)

>>> np.random.randint(10, size= 8)
array([5, 8, 3, 7, 6, 8, 5, 3])
```

```
# next we create an array with four cases and five variables

>>> np.random.randint(10, size=(4, 5))
array([[4, 7, 1, 9, 8],
       [7, 4, 0, 3, 7],
       [3, 6, 9, 3, 4],
       [4, 7, 3, 3, 9]])
```

```
# and then we create a three-dimensional array

>>> np.random.randint(10, size=(4, 5, 6))

array([[[7, 6, 1, 0, 1, 2],
        [3, 2, 3, 2, 9, 2],
        [5, 5, 6, 5, 0, 2],
        [1, 6, 2, 1, 9, 0],
        [1, 5, 3, 1, 3, 6]],

       [[0, 1, 9, 3, 4, 3],
        [2, 6, 6, 6, 6, 2],
        [2, 3, 7, 6, 2, 5],
        [1, 7, 7, 7, 0, 3],
        [3, 8, 6, 4, 1, 6]],

       [[1, 1, 2, 3, 2, 6],
        [0, 8, 1, 8, 1, 7],
        [7, 4, 3, 1, 6, 7],
        [7, 3, 1, 9, 8, 8],
        [9, 0, 1, 7, 2, 7]],

       [[3, 3, 2, 2, 5, 6],
        [4, 9, 3, 7, 4, 4],
        [8, 6, 3, 3, 7, 0],
        [9, 7, 0, 5, 2, 0],
        [3, 3, 9, 5, 2, 9]]])

# with random.rand() we generate real numbers; with random.
randint() we generate whole numbers
```

We can set a seed to be sure to generate the same random numbers and then repeat the examples:

```
>>> np.random.seed(12345)

>>> np.random.rand()
0.9296160928171479

>>> np.random.rand()
0.3163755545817859

>>> np.random.seed(12345)

>>> np.random.rand()
0.9296160928171479

>>> np.random.rand()
0.3163755545817859
```

We can generate random numbers using integers with random. randint():

```
>>> print(random.randint(0, 100))
27

>>> print(random.randint(0, 100))
32

# 0 and 100 are the limits within which we can extract
elements. In this case, the number 100 cannot be extracted.
This means that if we want to simulate, for example,
crapshooting, we set the limits between 1 and 7.
```

We can also create an object and extract the elements randomly:

```
>>> test1 = ["object1", "object2", "object3", "object4",
"object5"]
```

```
>>> print(random.choice(test1))
object2
```

```
>>> print(random.choice(test1))
object4
```

We can create another random object using np.random.randn(), which generates a normal distribution.

```
>>> x = np.random.randn(1000)
```

```
# we import matplotlib (described in Chapter 10)
```

```
>>> import matplotlib.pyplot as plt
```

```
# and create a histogram of the x object
```

```
>>> plt.hist(x, bins = 100)
```

```
# we display the histogram
```

```
>>> plt.show()
```

NumPy can also be used to generate random datasets, like the one plotted in Figure 9-1. But, we must also load the pandas package. We've already loaded the NumPy package, so let's proceed as follows:

```
# we import pandas
```

```
>>> import pandas as pd
```

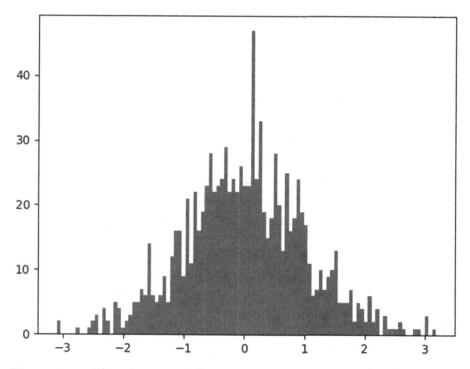

Figure 9-1. *Plot of a casual dataset created with NumPy and pandas*

We create a data frame using the pandas DataFrame() function. We identify a function immediately as belonging to a package, because the function is a method of that package:

```
package_name.function_name()
```

```
# we use the DataFrame () function of the pandas (pd) package
to create the datagram, and the random.randn() function of the
Numpy package (np) to generate random data. In brackets, we
first put the number of cases to be generated (in this case,
10) followed by the number of variables (in this case, 5)
```

```
>>> rdf = pd.DataFrame(np.random.randn(10,5))
```

```
>>> rdf
```

	0	1	2	3	4
0	0.669980	0.626433	-0.693932	-0.841258	-0.165200
1	0.108567	-0.743791	0.367369	0.645242	-0.297283
2	1.674781	0.241534	-0.403371	0.175751	0.274626
3	-2.339962	-0.083003	-1.387095	1.559257	-1.025012
4	0.383104	0.968755	0.236508	0.186294	0.094319
5	0.956150	-1.366423	0.694575	-0.107877	1.727657
6	-0.699931	-1.184346	0.581632	0.333015	-1.137382
7	0.867757	-0.872935	0.417772	-0.045722	0.432780
8	-0.685488	1.046816	0.465459	-0.446164	0.227635
9	-0.019854	0.643384	1.459784	0.559970	-0.358676

If we recreate a data frame using the same instructions, we will almost certainly get different data. To get the same data, we need to use a function that allows us to set a seed. In this way, we replicate the data.

```
# we set the seed
>>> np.random.seed(12345)
# we create the data frame
>>> rdf = pd.DataFrame(np.random.randn(10,5))
# we visualize the data frame
rdf
```

	0	1	2	3	4
0	-0.204708	0.478943	-0.519439	-0.555730	1.965781
1	1.393406	0.092908	0.281746	0.769023	1.246435
2	1.007189	-1.296221	0.274992	0.228913	1.352917
3	0.886429	-2.001637	-0.371843	1.669025	-0.438570
4	-0.539741	0.476985	3.248944	-1.021228	-0.577087
5	0.124121	0.302614	0.523772	0.000940	1.343810
6	-0.713544	-0.831154	-2.370232	-1.860761	-0.860757
7	0.560145	-1.265934	0.119827	-1.063512	0.332883
8	-2.359419	-0.199543	-1.541996	-0.970736	-1.307030
9	0.286350	0.377984	-0.753887	0.331286	1.349742

```
# we reinsert the same seed
>>> np.random.seed(12345)
# we recreate a dataset
>>> rdf2 = pd.DataFrame(np.random.randn(10,5))
# the generated data are identical
>>> rdf2
          0         1         2         3         4
0 -0.204708  0.478943 -0.519439 -0.555730  1.965781
1  1.393406  0.092908  0.281746  0.769023  1.246435
2  1.007189 -1.296221  0.274992  0.228913  1.352917
3  0.886429 -2.001637 -0.371843  1.669025 -0.438570
4 -0.539741  0.476985  3.248944 -1.021228 -0.577087
5  0.124121  0.302614  0.523772  0.000940  1.343810
6 -0.713544 -0.831154 -2.370232 -1.860761 -0.860757
7  0.560145 -1.265934  0.119827 -1.063512  0.332883
8 -2.359419 -0.199543 -1.541996 -0.970736 -1.307030
9  0.286350  0.377984 -0.753887  0.331286  1.349742
>>> np.random.seed(12345)
>>> rdf = pd.DataFrame(np.random.rand(10,5))
>>> rdf
          0         1         2         3         4
0 -0.204708  0.478943 -0.519439 -0.555730  1.965781
1  1.393406  0.092908  0.281746  0.769023  1.246435
2  1.007189 -1.296221  0.274992  0.228913  1.352917
3  0.886429 -2.001637 -0.371843  1.669025 -0.438570
4 -0.539741  0.476985  3.248944 -1.021228 -0.577087
5  0.124121  0.302614  0.523772  0.000940  1.343810
6 -0.713544 -0.831154 -2.370232 -1.860761 -0.860757
7  0.560145 -1.265934  0.119827 -1.063512  0.332883
8 -2.359419 -0.199543 -1.541996 -0.970736 -1.307030
9  0.286350  0.377984 -0.753887  0.331286  1.349742
```

Note that our completely random variables do not have a name; they are identified merely by numbers. Let's change the column names:

```
# we create a list that contains the names of the variables
>>> var_names = ['var1', 'var2', 'var3', 'var4', 'var5']
# we use the .columns method
>>> rdf2.columns = var_names
# we check the first cases of the data frame
>>> rdf2.head(2)
        var1      var2      var3      var4      var5
0 -0.204708  0.478943 -0.519439 -0.555730  1.965781
1  1.393406  0.092908  0.281746  0.769023  1.246435
# the variable names are correct
```

We can acquire other types of distributions using other NumPy functions.

```
# binomial distribution
>>> rdf_bin = pd.DataFrame(np.random.binomial(100, 0.5, (10,5)))
>>> rdf_bin
    0   1   2   3   4
0  47  56  48  42  53
1  43  56  51  56  46
2  50  42  40  55  46
3  46  43  54  51  53
4  41  48  47  45  42
5  53  55  51  50  58
6  51  57  46  53  48
7  56  53  46  54  54
8  53  50  52  53  46
9  50  46  54  57  56

# Poisson distribution
```

```
>>> rdf_poi = pd.DataFrame(np.random.poisson(100, (10,5)))

>>> rdf_poi
     0    1    2    3    4
0  109  107   98  111   95
1  115  109  101  108   95
2  105   97  102  100   94
3   94   93   94  122   96
4  117   85  135   90   83
5  103  106  105   93  116
6  111   95  100   95   80
7   81   75   84   93  101
8  105  109  104  104  113
9   97  120   90   98   95

# uniform distribution

>>> rdf_un = pd.DataFrame(np.random.uniform(1, 100,(10,5)))

>>> rdf_un
           0          1          2          3          4
0  49.139046  98.433411  60.777590  76.202858  68.767153
1  93.309206  95.028762  99.021002  13.489383  97.683461
2  23.681460  19.419586  51.411931  53.519271  28.981285
3  31.705992  31.678553  27.372500  59.749684  22.496303
4  40.835909  29.567563  18.210241  23.345639  98.875500
5  43.971084  82.990738  57.678770  65.538128  73.244077
6  39.737081  39.893383  86.095572  81.191942  83.817845
7  19.610805  36.600078  48.716414  96.641192  56.768005
8  82.922981   8.534653  55.760657  55.246106  90.638916
9  37.579379  90.215102  14.922471  10.818199  97.345143
```

How do we save a data frame or array created in NumPy? First, we create a data frame

```
>>> tos = pd.DataFrame(np.random.randn(10,5))
```

then save it with the NumPy save() function.

```
>>> np.save('tos_saved', tos)
# we can check it in our work directory
```

To load our saved file, we use the load() function in NumPy.

```
>>> load1 = np.load('tos_saved.npy')
```

NumPy contains other advanced mathematics modules, such as numPy.linalg for linear algebra, fft for Fourier transform, and a number functions for financial analysis, such as interest and futures calculations. You can find all the necessary documentation about NumPy features at https://docs.scipy.org/doc/.

Summary

NumPy and SciPy are two very important packages used by data analysts. Each has a variety of features, but the packages can be combined to create random datasets, for example.

CHAPTER 10

Matplotlib

Creating graphs is an important step in exploratory analysis, and one of the first stages in data analysis. We can use Matplotlib to construct a variety of analytical graphs that display our data in different ways.

Basic Plots

To represent our data graphically, we can use the Matplotlib library, one of the most used graphics representation software packages. Its documentation can be found at `https://matplotlib.org`. The section gallery of the Matplotlib site (at `https://matplotlib.org/gallery.html`) features a series of examples of charts with code.

Matplotlib is installed with Anaconda, so we have to import it.

```
# we import the Matplotlib library

>>> import matplotlib as mlp

>>> import matplotlib.pyplot as plt

>>> %matplotlib inline

# this last line of code allows us to view the charts directly
using Jupyter
```

© Valentina Porcu 2018
V. Porcu, *Python for Data Mining Quick Syntax Reference*,
https://doi.org/10.1007/978-1-4842-4113-4_10

We can represent elements by inserting them directly into the function, in the form of a list. The following code produces the plot in the Figure 10-1.

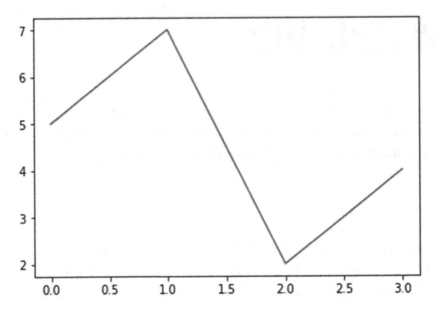

Figure 10-1. *A plotted list*

```
>>> plt.plot([5,7,2,4])
>>> plt.plot([5,7,2,4], [4,6,9,2], 'ro')
# 'ro' stands for round object
```

The plot of round objects is shown in Figure 10-2.

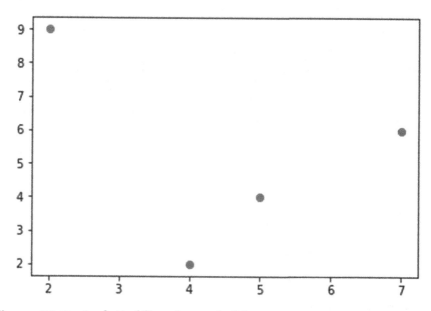

Figure 10-2. *A plotted list of round objects*

As shown in Figure 10-1, we can create a line. But, we can also customize color and type of representation by modifying arguments.

Let's can create two objects and represent them.

```
# we create two objects

>>> x = [ 50, 70, 90, 65]
>>> y = [129, 192, 163, 172]

>>> plt.plot(x, y, linewidth = 4.0)
```

The plot is shown in Figure 10-3.

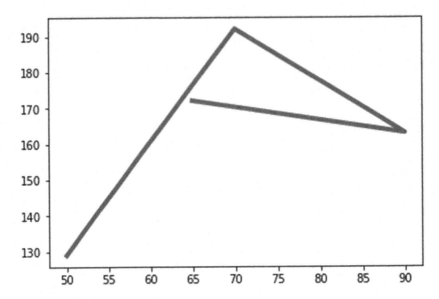

Figure 10-3. *A custom plot*

As you can see, we changed the line thickness with the line width argument. We can also modify the line with the **linestyle** argument, or **ls**:

```
>>> plt.plot(x, y, linewidth = 2.0, linestyle = '--')
```

The plot is shown in Figure 10-4.

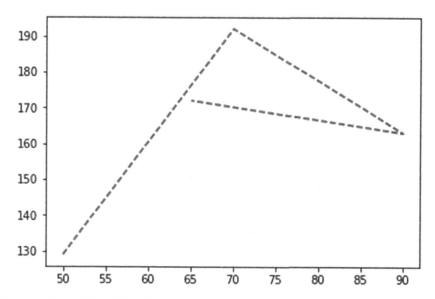

Figure 10-4. *Modified line style*

We can add markers to highlight data better:

```
>>> plt.plot(x, y, linewidth = 1.0, ls = '-', marker = "o",
markersize = 10)
```

The plot is shown in Figure 10-5.

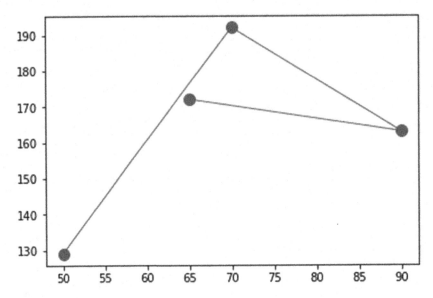

Figure 10-5. *Plot with markers*

We can customize the markers even more further by editing, for example, the inner color:

```
>>> plt.plot(x, y, linewidth = 1.0, ls = '-', marker = "o",
markersize = 10, markerfacecolor = 'white')
```

The plot is shown in Figure 10-6.

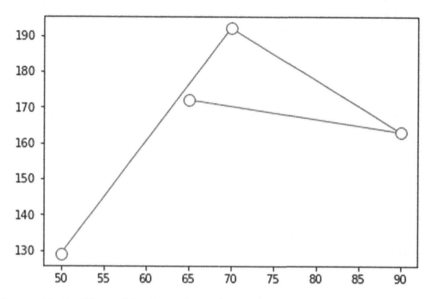

Figure 10-6. *Plot with altered marker color*

More information on how to customize a chart can be found using

```
>>> help(plt.plot)
```

Next we can add parameters to add a title and axes names.

```
>>> plt.plot(x, y)
>>> plt.title("TITLE")
>>> plt.xlabel("Axis X")
>>> plt.ylabel("Axis Y")
```

The plot is shown in Figure 10-7.

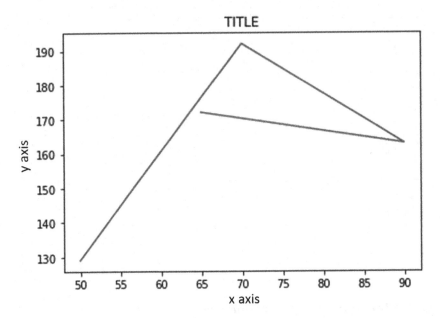

Figure 10-7. *Plot with a title and axes labels*

We can customize a chart even further by changing colors for all chart elements:

```
>>> plt.plot(x, y, color = "yellow")
>>> plt.title("TITLE", color = "blue")
>>> plt.xlabel("Axis X", color = "purple")
>>> plt.ylabel("Axis Y", color = "green")
```

The plot is shown in Figure 10-8.

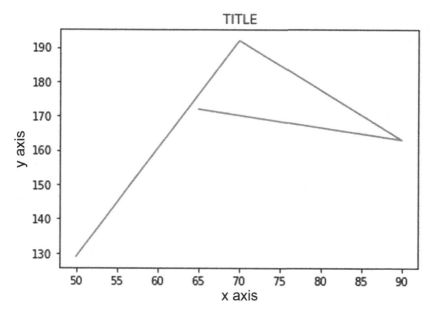

Figure 10-8. *Altering plot element colors*

We can also add a grid by using the 'grid' parameter, and a legend by using the 'legend' parameter.

```
>>> plt.plot(x, y)
>>> plt.title("TITLE", color = "blue")
>>> plt.xlabel("Axis X", color = "purple")
>>> plt.ylabel("Axis Y", color = "green")
>>> plt.grid(True)
>>> plt.legend(['Legend1'])
```

The plot is shown in Figure 10-9.

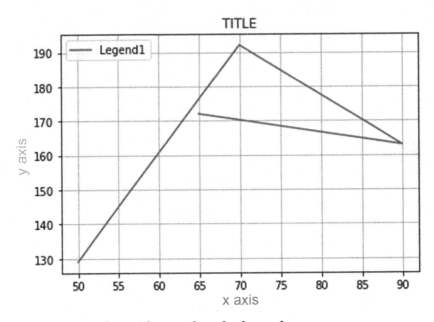

Figure 10-9. *A plot with a grid and a legend*

We can move the legend into the chart by changing the 'loc' parameter:

```
>>> plt.plot(x, y)
>>> plt.title("TITLE", color = "blue")
>>> plt.xlabel("Axis X", color = "purple")
>>> plt.ylabel("Axis Y", color = "green")
>>> plt.grid(True)
>>> plt.legend(['Legend2'], loc = 2)
```

The plot is shown in Figure 10-10.

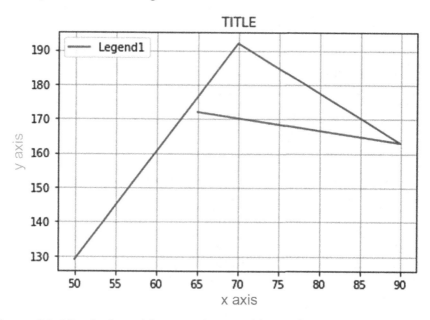

Figure 10-10. *A plot with a repositioned legend*

These are the possible positions of the legend:

0

1 = top right

2 = top left

3 = bottom right

4 = lower left

5 = to the right

6 = centered left

7 = centered right

8 = centered low

9 = centered high

10 = centered

The codes for color are as follows:

    b = blue

    c = cyan

    g = green

    m = magenta

    r = red

    y = yellow

    k = black

    w = white

We can also change the shapes used in a plot:

```
>>> plt.plot([1,2,3,4],[1,4,8,15],'b*')
>>> plt.plot([1,3,5,7],[1,4,8,12],'g^')
>>> plt.plot([1,2,3,5],[2,5,4,12],'ro')
>>> plt.legend(['First','Second','Third'],loc=0)
```

The plot is shown in Figure 10-11.

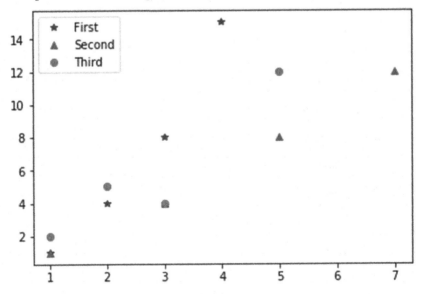

Figure 10-11. *A plot with points of different shape*

Now let's create subcharts using the subplot() function:

```
>>> plt.subplot(2,2,1)
>>> plt.plot([1,2,3,4],[1,4,8,15],'b*')

>>> plt.subplot(2,2,2)
>>> plt.plot([1,3,5,7],[1,4,8,12],'g^')

>>> plt.subplot(2,2,3)
>>> plt.plot([1,2,3,5],[2,5,4,12],'ro')

>>> plt.subplot(2,2,4)
>>> plt.plot([1,2,3,5],[2,5,4,12],'b')
```

The plot is shown in Figure 10-12.

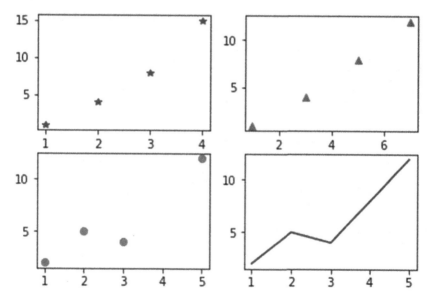

Figure 10-12. *Creation of subplots*

213

We can indicate how many charts we want (in this case, two) and how they are placed (in this case, side by side):

```
>>> plt.subplot(1,2,1)
>>> plt.plot([1,2,3,4],[1,4,8,15],'b*')

>>> plt.subplot(1,2,2)
>>> plt.plot([1,3,5,7],[1,4,8,12],'g^')
```

The plot is shown in Figure 10-13.

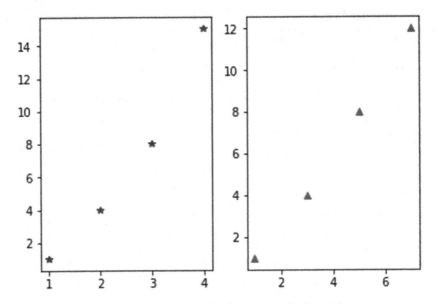

Figure 10-13. *Creation of two subplots set side by side*

Pie Charts

Now let's see how to create pie charts: a pie chart can be used to show the composition of something (like a market). To plot a pie we can use the `plt.pie()` function:

```
>>> plt.pie(x)
```

The plot is shown in Figure 10-14.

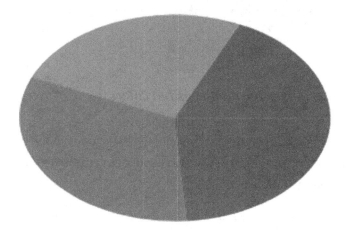

Figure 10-14. *A basic pie chart*

We can customize it by editing its colors:

```
# we create a palette of colors
>>> col1 = ["yellow", "red", "purple", "orange"]
# we apply the new colors to the chart
>>> plt.pie(x, colors = col1)
```

The plot is shown in Figure 10-15.

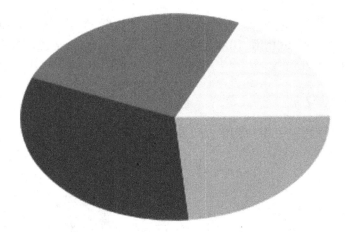

Figure 10-15. *A pie chart with custom colors*

To modify colors even further, we can use hex codes. A list of the codes can be found at http://cloford.com/resources/colours/500col.htm.

Let's add some labels:

```
>>> lab1 = ['A','B','C','D']
>>> plt.pie(x, colors = col1, labels = lab1)
```

The plot is shown in Figure 10-16.

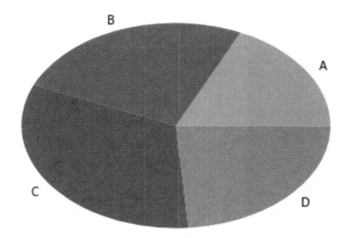

Figure 10-16. *A pie chart with labels*

We can separate sections of the pie by using the 'explode' parameter. We can even indicate the distance among the exploded pie sections:

```
>>> ex1 = [0.5,0,0,1]
>>> lab1 = ['A','B','C','D']
>>> plt.pie(x, colors = col1, labels = lab1, explode = ex1)
```

The plot is shown in Figure 10-17.

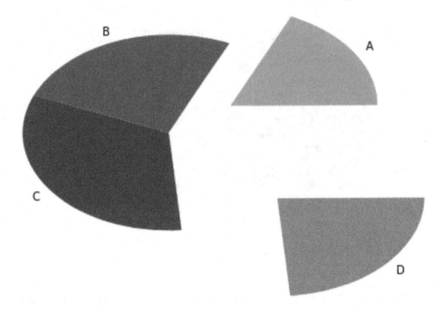

Figure 10-17. An exploded pie chart

Other Plots and Charts

We can create yet other types of plots and charts. For example, we can build a scatterplot. A scatterplot is very useful to see the relationship between two variables.

```
>>> plt.scatter(x, y)
```

The plot is shown in Figure 10-18.

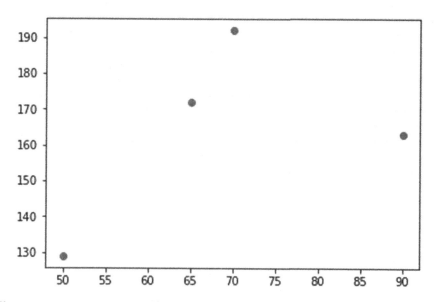

Figure 10-18. *A scatterplot*

We can create bar charts with the plt.bar() function. Bar charts and histograms are very useful to compare our data and also to display categorical variables:

```
>>> plt.bar(x, y)
```

The plot is shown in Figure 10-19.

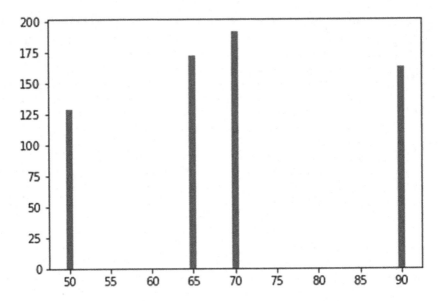

Figure 10-19. *A bar chart*

We can change the orientation of a bar chart:

```
>>> plt.barh(x, y)
```

The plot is shown in Figure 10-20.

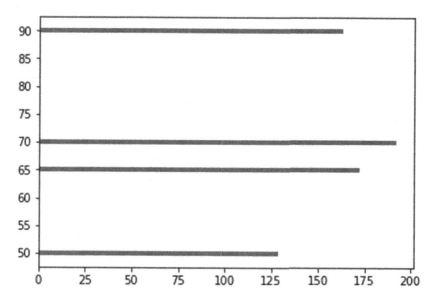

Figure 10-20. *A reoriented bar chart*

We can create a chart from a data frame. To do this, we must import pandas for dataset and NumPy management. Let's generate a random set of ten cases and four variables.

```
>>> import pandas as pd
>>> import numpy as np

>>> df1 = pd.DataFrame(np.random.rand(10, 4), columns =
['var1', 'var2', 'var3', 'var4'])

>>> df1.plot(kind = "bar")
```

221

The plot is shown in Figure 10-21.

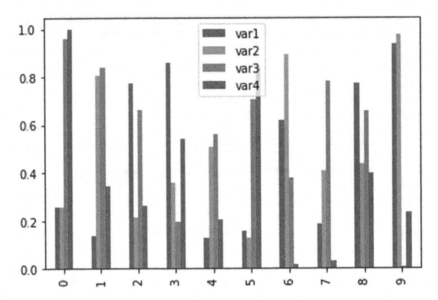

Figure 10-21. *A bar chart created using a random data frame*

To create stacked bars, we use the parameter 'stacked':

```
>>> df1.plot(kind = "bar", stacked = True)
```

The plot is shown in Figure 10-22.

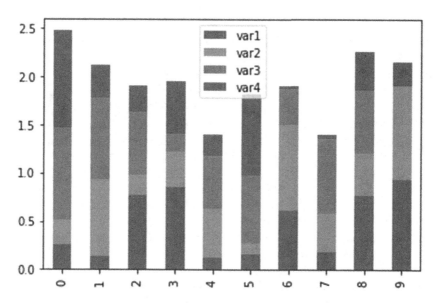

Figure 10-22. *A chart with stacked bars*

We can create a histogram that represents the variables of the dataset. (Histograms are discussed in more detail at the end of the chapter.)

```
>>> df1.hist()
```

The plot is shown in Figure 10-23.

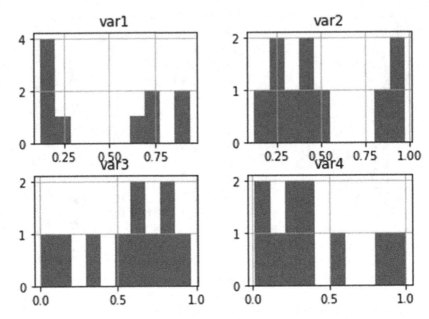

Figure 10-23. *Multiple histograms for each variable in the dataset*

Or, we can represent a single variable:

```
>>> df1['var1'].hist()
```

The plot is shown in Figure 10-24.

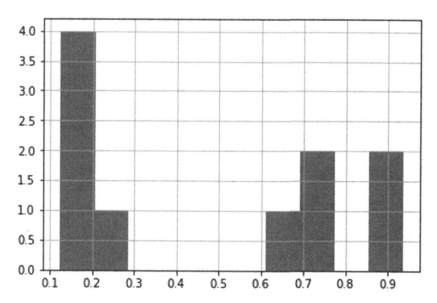

Figure 10-24. *A histogram of one variable*

We can also select a column using methods other than the name, such as the .loc method.

```
>>> df1.loc[1].hist()
```

We create box plots by using the boxplot() function. This kind of visualization can be used to show the shape of the distribution, its central value, and its variability:

```
>>> df1.boxplot(return_type = "axes")
```

The plot is shown in Figure 10-25.

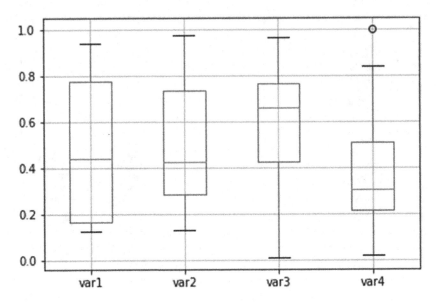

Figure 10-25. *A boxplot*

We can build area charts:

```
>>> df1.plot(kind = "area")
```

The plot is shown in Figure 10-26.

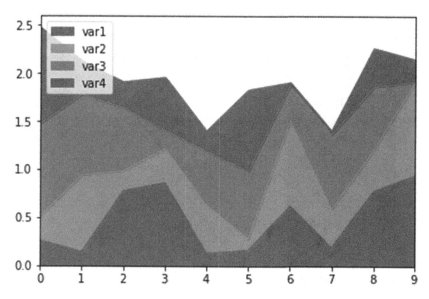

Figure 10-26. *An area chart*

Each function that we use has its own parameters, which we can change, as we saw in the first section of this chapter. For instance, we can change the colors of the area chart by applying the palette we already created:

```
>>> df1.plot(kind = "area", color = col1)
```

The plot is shown in Figure 10-27.

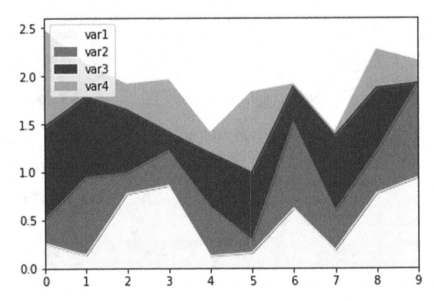

Figure 10-27. *An area chart with an altered color palette*

Saving Plots and Charts

We can save our plots and charts with the .savefig method. We can also designate its name and set the resolution (dots per inch) as well:

```
>>> df1.plot(kind = "scatter", x = "var3", y = "var4")
# we save the image in the working directory in the following way
>>> plt.savefig('graph1.png', dpi = 600)
```

Let's check whether the chart has been saved successfully to our working directory. We can use the image downloaded for example for a presentation, or including it in a report after the data analysis or to explain our data in an exploratory phase.

Selecting Plot and Chart Styles

Matplotlib also includes a set of styles that can be applied to charts. We can view these styles by typing:

```
>>> plt.style.available
```

```
['bmh',
 'classic',
 'dark_background',
 'fivethirtyeight',
 'ggplot',
 'grayscale',
 'seaborn-bright',
 'seaborn-colorblind',
 'seaborn-dark-palette',
 'seaborn-dark',
 'seaborn-darkgrid',
 'seaborn-deep',
 'seaborn-muted',
 'seaborn-notebook',
 'seaborn-paper',
 'seaborn-pastel',
 'seaborn-poster',
 'seaborn-talk',
 'seaborn-ticks',
 'seaborn-white',
 'seaborn-whitegrid',
 'seaborn']
```

To apply a style, we must insert a line of code that features the theme name:

```
>>> plt.style.use('dark_background')
>>> df1.plot(kind = "area")
```

The plot is shown in Figure 10-28.

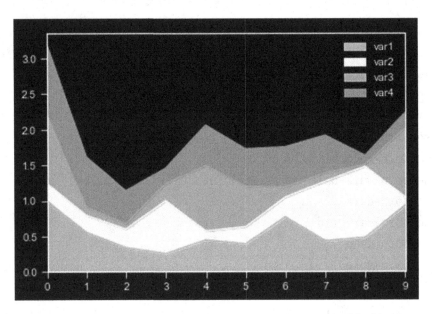

Figure 10-28. *A custom area chart that uses a Matplotlib "dark background" theme*

Here is another example:

```
>>> plt.style.use('seaborn-darkgrid')
>>> df1.plot(kind = "area")
```

The plot is shown in Figure 10-29.

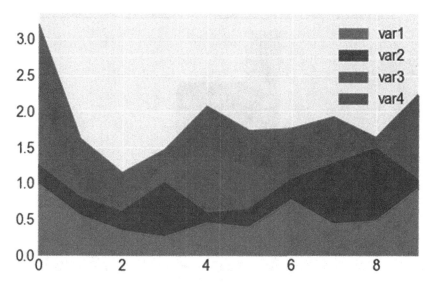

Figure 10-29. *A custom area chart with a "seaborn" theme*

More on Histograms

We can create two random objects with NumPy and represent them graphically separately, then compile their data into one chart:

```
>>> df2 = np.random.randn(100)
>>> df3 = np.random.randn(100)

>>> plt.hist(df2)
```

The first plot is shown in Figure 10-30.

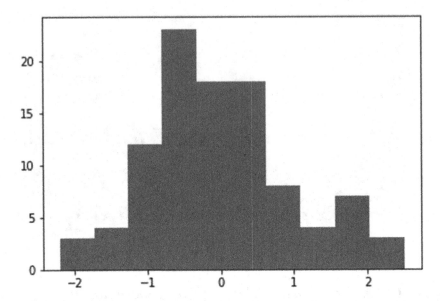

Figure 10-30. *The first histogram*

Now let's display the second histogram.

```
>>> plt.hist(df3)
```

The plot is shown in Figure 10-31.

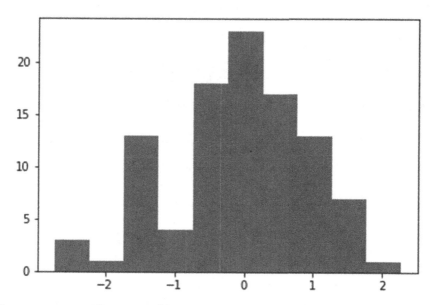

Figure 10-31. *The second histogram*

Now let's combine the two datasets:

```
>>> plt.hist(df2, color = "red", alpha = 0.3, bins = 15)
>>> plt.hist(df3, alpha = 0.6, bins = 15)
```

```
# we present the two datasets together and define whether we
want color, transparency through the alpha parameter, and the
number of intervals into which we want data to be divided.
```

The plot is shown in Figure 10-32.

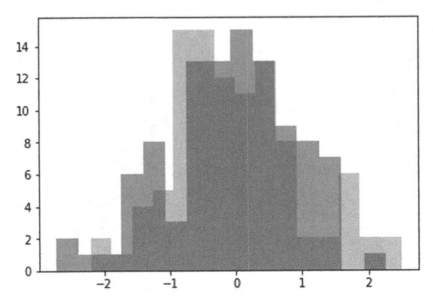

Figure 10-32. *A combined histogram*

Matplotlib is just one of many Python packages that can be used to display data. Other chart creation packages can be found at http:// pbpython.com/visualization-tools-1.html. One of the most used data mining charts, for example, is seaborn.

Summary

Matplotlib is one of the most basic libraries for plotting data. Plotting datasets for data analysis is crucial to understanding the relationships among variables.

CHAPTER 11

Scikit-learn

Scikit-learn is one of the most important and most used packages for machine learning with Python. It features many functions for various predictive algorithms. In this chapter, we examine some of the algorithms included in the Scikit-learn package. Given the breadth of the subject, the example presented reflect the most used models. Those of you who have no prior knowledge of machine learning may find it difficult to understand some of the techniques presented in this chapter, which are not explained in detail.

What Is Machine Learning?

Machine learning is a branch of data analysis that transforms datasets built in a particular way into predictions that can be applied to new data. Machine learning uses data we already have to predict future behaviors. Machine-learning techniques have been a real revolution in data mining and they have a great impact on a variety of fields of application.

Machine learning is widespread among many applications used every day. Large companies such as Amazon, Netflix, Google, Apple, and Facebook use machine-learning algorithms for various reasons. For instance, Facebook uses machine learning to recognize faces in images; Amazon and Netflix analyze customer preferences (the last thing you viewed or bought) to propose new products that might match your interests.

© Valentina Porcu 2018
V. Porcu, *Python for Data Mining Quick Syntax Reference*,
https://doi.org/10.1007/978-1-4842-4113-4_11

Google, for example, uses machine learning in translation and automatic driving, or also suggesting the road with less traffic based on our habits or the place where we are used to go on a given day of the week. Machine learning also helps us detect spam messages from non spam messages, often using probabilistic methods or by combining multiple methods (for instance, adding probabilistic methods to keywords and user-defined rules).

Apple and Microsoft use machine learning to provide us with a voice assistant that helps us work with a phone or tablet using only our voice. Other companies are currently refining artificial intelligence methods, automated driving, and more.

The field of machine learning gained wider attention when a supercomputer (Watson), developed by IBM, took part to the *Jeopardy!* quiz program.

Machine learning has also been used to predict election results, first by Nate Silver, a scientist who, in October 2012, published a preview of US elections, and whose results were very close to the actual data.

Predictive data mining is used in the healthcare field. Patient data and clinical records can help identify people who are at greater risk of contracting certain conditions and illness such as diabetes or heart disease. DNA analysis and genetic kits have been used, for example, to detect genes responsible for or otherwise related to certain types of cancer, including breast cancer.

One of the most outdated uses of machine learning is handwriting recognition—in particular, handwritten addresses and zip codes. Recognition is based on various occurrences of each handwritten number using neural networks (conducted by Bell Labs).

Research in machine learning and related topics, such as deep learning and artificial intelligence, improves every day and is at the forefront of the computing world. Some web sites such as Kaggle publish contents every few days during which subscribers try to solve a given problem. The most famous Kaggle contest was announced by Netflix. In 2006, Netflix awarded a $1,000,000 USD prize for a recommendation system. The system that was designed has not been implemented by Netflix because it is too complex and computationally expensive.

Let's look at the various modules and techniques in the Scikit-learn package.

Import Datasets Included in Scikit-learn

First, the scikit-learn package includes some datasets, which we can import like this:

```
>>> from sklearn import datasets
```

To import one of the datasets, type

```
>>> iris = datasets.load_iris()
```

The iris dataset is made up of petals and sepals of three different types of iris: *versicolor*, *virginica*, and *setosa*. It contains 150 equally divided cases on three types of flowers, and five variables.

In the previous chapters we saw how to import a dataset in .csv from an external file on our local computer or from a website using a link. Scikit-learn also includes some of the most used dataset for data mining, like iris and Boston. The format used in Scikit-learn can result a little bit confusing for a beginner. All the dataset is included in a same object, but not in a tabular format like the .csv files we examined in the previous chapters. Datasets in Scikit-learn contain data in array format, the target or label (the variable we want to predict) and some other informations, like variable description (in the DESCR object) and the columns names in two distinct objects: feature_names for the variables, and target_names for the target or label.

```
>>> type(iris)
sklearn.datasets.base.Bunch
```

```
>>> iris.keys()
dict_keys(['data', 'target', 'target_names', 'DESCR',
'feature_names'])
```

We can display the actual dataset:

```
>>> iris.data
array([[ 5.1,  3.5,  1.4,  0.2],
       [ 4.9,  3. ,  1.4,  0.2],
       [ 4.7,  3.2,  1.3,  0.2],
       [ 4.6,  3.1,  1.5,  0.2],
       [ 5. ,  3.6,  1.4,  0.2],
       [ 5.4,  3.9,  1.7,  0.4],
       [ 4.6,  3.4,  1.4,  0.3],
       [ 5. ,  3.4,  1.5,  0.2],
       [ 4.4,  2.9,  1.4,  0.2],
[...]
```

In this way, we display only the numeric data. To see the actual classification, we must proceed as follows:

```
>>> iris.target
```

```
array([0, 0, 0, 0, 0, 0, 0, 0, 0, 0, 0, 0, 0, 0, 0, 0, 0, 0, 0, 0, 0, 0, 0,
       0, 0, 0, 0, 0, 0, 0, 0, 0, 0, 0, 0, 0, 0, 0, 0, 0, 0, 0, 0, 0, 0,
       0, 0, 0, 0, 1, 1, 1, 1, 1, 1, 1, 1, 1, 1, 1, 1, 1, 1, 1, 1, 1, 1,
       1, 1, 1, 1, 1, 1, 1, 1, 1, 1, 1, 1, 1, 1, 1, 1, 1, 1, 1, 1, 1, 1,
       1, 1, 1, 1, 1, 1, 1, 1, 2, 2, 2, 2, 2, 2, 2, 2, 2, 2, 2, 2, 2, 2,
       2, 2, 2, 2, 2, 2, 2, 2, 2, 2, 2, 2, 2, 2, 2, 2, 2, 2, 2, 2, 2, 2,
       2, 2, 2, 2, 2, 2, 2, 2, 2, 2, 2, 2])
```

```
# data target is numeric because scikit-learn does not accept
categorical data by default, so it is necessary to encode data
in numeric form
```

```
# we display the names
```

```
>>> iris.target_names
```

```
array(['setosa', 'versicolor', 'virginica'],
      dtype='<U10')
```

We can display the number of cases and variables as follows:

```
>>> iris.data.shape
```

```
(150, 4)
```

We can acquire a description of data by using .DESCR:

```
>>> iris.DESCR
```

```
{'DESCR': 'Iris Plants Database\n====================\n\nNotes\n-----\nData Set Characteristics:\n    :Number of Inst
ances: 150 (50 in each of three classes)\n    :Number of Attributes: 4 numeric, predictive attributes and the class\n
:Attribute Information:\n        - sepal length in cm\n        - sepal width in cm\n        - petal length in cm\n
- petal width in cm\n        - class:\n            - Iris-Setosa\n            - Iris-Versicolour\n
- Iris-Virginica\n    :Summary Statistics:\n\n    ==============  ====  ====  =======  =====  ====================\n
Min  Max   Mean    SD   Class Correlation\n    ==============  ====  ====  =======  =====  ====================\n    sepal
length:   4.3  7.9   5.84   0.83    0.7826\n    sepal width:    2.0  4.4   3.05   0.43   -0.4194\n    petal length:
1.0  6.9   3.76   1.76    0.9490  (high!)\n    petal width:    0.1  2.5   1.20   0.76    0.9565  (high!)\n    =======
==============  ====  ====  =======  =====  ====================\n\n    :Missing Attribute Values: None\n    :Class Distribution:
33.3% for each of 3 classes.\n    :Creator: R.A. Fisher\n    :Donor: Michael Marshall (MARSHALL%PLU@io.arc.nasa.gov)\
```

Creation of Training and Testing Datasets

In machine learning we tipically start from a dataset in a .csv format. From
this file we will create 4 pieces. One thing that Scikit-learn allows us to do
is to create a training dataset and a testing dataset. In machine learning,
we typically start with a labeled dataset and divide it into two parts: one
for training (about 70%–80% of the dataset), which is used to train the
algorithm; and one for testing (the remaining 20%–30%), which is used to
test the efficacy of the data algorithm. This feature allows us to compare
actual data with those predicted by the algorithm and see how they work.
Training and testing datasets are also divised in two parts: one with the
variables that we will use to create a model, and one with the variable we
want to learn to predict.

```
>>> from sklearn.model_selection import train_test_split
```

```
>>> x_train, x_test, y_train, y_test = train_test_split(x, y,
test_size = 0.3)
```

```
# we only need to specify the percentage of the test dataset—in
this case, 30% (0.3)
```

If we apply this to our iris dataset, for example, we can create four objects: a training object that contains four variables of the iris dataset and 70% of the cases, a test object that contains the rest of the elements (30%), and a label or target variable, which is always divided in two.

```
>>> x_train, x_test, y_train, y_test = train_test_split(iris.
data, iris.target, test_size = 0.3)
```

Let's check the size of the various objects created:

```
>>> x_train.data.shape
(105, 4)
```

```
>>> x_test.data.shape
(45, 4)
```

```
>>> y_train.data.shape
(105,)
```

```
>>> y_test.data.shape
(45,)
```

Preprocessing

Scikit-learn permits preprocessing of data (although we do not need to do so with the iris dataset).

```
>>> from sklearn import preprocessing
```

```
# iris_scaled = pd.DataFrame(preprocessing.scale(iris_data))
```

Regression

Regression analysis is used to explain the relationship between a variable, y, called a *response variable* or *dependent variable*, and one or more independent variables.

To calculate the regression, let's import the correct model from Scikit-learn:

```
>>> from sklearn.linear_model import LinearRegression

# we simplify the work a bit by creating a copy of the
regression model

>>> lr = LinearRegression()

# we create the model using the training objects

>>> lr.fit(x_train, y_train)

# we view the coefficients

>>> print(lr.intercept_)

>>> lr.coef_

# we predict the membership for the test objects

>>> pred = lr.predict(x_test)

>>> print(pred)
```

Now let's look at the code that allows us to apply metrics to measure model efficacy:

```
>>> from sklearn import metrics

>>> print('MAE', metrics.mean_absolute_error(y_test, pred))
>>> print('MSE', metrics.mean_squared_error(y_test, pred))
>>> print('RMSE', np.sqrt(metrics.mean_squared_error(y_test,
    pred)))

>>> metrics.explained_variance_score(y_test, pred)
```

K-Nearest Neighbors

The k-nearest neighbor algorithm is a supervised algorithm used for data prediction and data mining. It is also used for pattern recognition (such as facial recognition), for identifying patterns in genetic code, for identifying illnesses, and for film and music recommendation systems. The logic behind the k-neighbor algorithm can be summed up in the Latin phrase *"Similes cum similar bus facillime congregantur"*—meaning, "similar ones gather together easily." In short, we use the algorithm to analyze cases in a dataset to find similar elements. New cases are then aggregated to newly formed groups, depending on how close they are to a group and how far from one another. The k-neighbor algorithm then calculates the distance of the unclassified item to the others, and assigns it the closest class of element (or elements, k). k is nothing more than the number of close observations that we can use to determine the class of an item with an unknown class. For example, if we set k equal to two, we assign the class to the item based on its two closest elements. If we set it equal to three, we get the three closest elements, and so on.

```
# we cannot use the k-neighbor algorithm on the iris dataset,
so from this point onward, we limit ourselves to giving an idea
of the code for the various classification models

>>> from sklearn.neighbors import KNeighborsClassifier
>>> knn = KNeighborsClassifier(n_neighbors = 3)

>>> x_train = from the dataset we will use the variables except
    the label
>>> y_train = the test labels
>>> x_test = new data
>>> y_test = new data labels

>>> knn.fit(x_train, y_train)
>>> new_pred = knn.predict(x_new)
>>> print(new_pred)
```

When dealing with classification, we use methods other than those of regression to test the adequacy of a model:

```
>>> from sklearn.metrics import classification_report,
confusion_matrix

>>> print(confusion_matrix(y_test, y_pred))

>>> print(classification_report(y_test, y_pred))
```

Cross-validation

Cross-validation consists of dividing a dataset into a number of equal parts generally indicated by k (often five or ten parts) then testing the adequacy of the prediction model on these k groups.

```
>>> from sklearn.model_selection import cross_val_score
>>> cv5 = cross_val_score(model, x_train, y_train, cv = 5)
>>> cv10 = cross_val_score(model, x_train, y_train, cv = 10)
>>> print(np.mean(cv5))
>>> print(np.mean(cv10))
```

Support Vector Machine

Support Vector Machine (SVM) is used to determine the boundary between items belonging to two different classes, then projecting them into multidimensional space to discern the hyperplane that maximizes margins between the two sets of data.

```
>>> from sklearn.svm import SVC
>>> clf = svm.SVC(kernel='linear', C=1).fit(x_train, y_train)

>>> clf.score(x_test, y_test)
>>> from sklearn.model_selection import cross_val_predict
>>> pred = cross_val_predict(clf, iris.data, iris.target, cv=10)
>>> metrics.accuracy_score(iris.target, pred)
```

Decision Trees

The basic idea behind a decision tree is a *divide et impera* model, in which at each step we can reduce variability between nodes. Let's us start with the entire dataset, which is then divided into smaller groups that are based more homogeneously and intrinsically on internal characteristics.

```
>>> from sklearn.tree import DecisionTreeClassifier
>>> dtc = DecisionTreeClassifier()
>>> dtc = dtc.fit(x_train, y_train)
>>> pred = dtc.predict(y_test)
>>> sklearn.metrics.confusion_matrix(y_test, pred)
>>> sklearn.metrics.accuracy_score(y_test, pred)
>>> sklearn.metrics.classification_report(y_test, pred)
```

KMeans

KMeans is an unsupervised method of classification, which means we do not have a label to guide us during classification. For this reason, we choose to use clustering as a helpful exploratory analysis method, because it allows us to group elements of a dataset based on how similar or dissimilar they are.

Clustering includes a set of methods that allows segmentation of a heterogeneous population into homogeneous subgroups. Of all the clustering methods available, KMeans is one of the important ones. The basic concept of clustering is based on the fact that we divide the items of a set into homogeneous groups without labeling them initially. Label-free data must be grouped in such a way that they are not only homogeneous within their clusters, but also heterogeneous to other elements of the other clusters. After we split our clustered items, we can classify new items as belonging to either cluster of the first dataset.

```
>>> from sklearn.cluster import KMeans
>>> kmeans = KMeans(n_clusters=4)
>>> kmeans.fit(df)
>>> pred = kmeans.predict(df)
>>> pred
```

This was just a cursory discussion of machine learning using the Scikit-learn package. Machine learning is a challenging topic and therefore not easy to sum up in a few pages. I thought it would be helpful to expose to some predictive data mining concepts and various Scikit-learn modules that can be used for machine learning.

Managing Dates

Managing dates using Python is important, especially when dealing with time series representations. We can handle dates using the datetime package and pandas. First, we must import datetime.

```
>>> import datetime as dt

# we create a first object that contains time

>>> t1 = dt.time(19, 43 , 30)

>>> print(t1)
19:43:30

# to create an object featuring a date, we use dates

>>> dt.date.today()

>>> datetime.date(2017, 3, 28)
```

```
# we can query the created object about the year, the month,
the day

>>> today = dt.date.today()

>>> today.year
2017

>>> today.month
3

>>> today.day
28

>>> t2 = dt.date(2016, 5, 20)

>>> print(t2)

2016-05-20

# we can query an object to find the year, month, and day

>>> t2.year
2016

>>> t2.month
5

>>> t2.day
20

# we can find the exact hour and minute from our computer

>>> dt.datetime.now()

>>> datetime.datetime(2017, 3, 30, 13, 4, 52, 591324)
```

Resources for parsing a date are available at http://strftime.org/.

Let's carry on with date management using pandas.

```
>>> import pandas as pd

# we can manage various data formats through Timestamp

>>> pd.Timestamp("2016-3-7")

>>> pd.Timestamp("2016/4/10")

>>> pd.Timestamp("2015, 12, 10")

>>> pd.Timestamp("2015, 12, 10 12:42:57")

>>> date1 = ["2016/4/10", "2015, 12, 10", "2015, 12, 10
    12:42:57"]

>>> print(date1)
['2016/4/10', '2015, 12, 10', '2015, 12, 10 12:42:57']

>>> type(date1)
list

>>> pd.to_datetime(date1)

DatetimeIndex(['2016-04-10 00:00:00', '2015-12-10 00:00:00',
               '2015-12-10 12:42:57'],
            dtype='datetime64[ns]', freq=None)

# we create another object that contains our dates, but also
some other element

>>> date2 = ["2016/4/10", "2015, 12, 10", "2015, 12, 10
12:42:57", "October", "2011", "test"]

# if we pass this object in Timestamp, we get an error

>>> pd.to_datetime(date2)
```

```
# we can handle errors with the 'coerce' parameter

>>> pd.to_datetime(date2, errors = "coerce")

DatetimeIndex(['2016-04-10 00:00:00', '2015-12-10 00:00:00',
               '2015-12-10 12:42:57',                  'NaT',
               '2011-01-01 00:00:00',                  'NaT'],
              dtype='datetime64[ns]', freq=None)
```

Dates that are not recognized are identified as NaT.

Let's carry on and create a range of dates:

```
>>> period1 = pd.date_range(start = "2016   01 01", end = "2016
    12 31")

>>> print(period1)

DatetimeIndex(['2016-01-01', '2016-01-02', '2016-01-03', '2016-01-04',
               '2016-01-05', '2016-01-06', '2016-01-07', '2016-01-08',
               '2016-01-09', '2016-01-10',
               ...
               '2016-12-22', '2016-12-23', '2016-12-24', '2016-12-25',
               '2016-12-26', '2016-12-27', '2016-12-28', '2016-12-29',
               '2016-12-30', '2016-12-31'],
              dtype='datetime64[ns]', length=366, freq='D')
```

The frequency argument (**freq='D'**) means that a day (day) interval is set, but we can modify it:

```
# for example, by inserting ten days

>>> pd.date_range(start = "2016   01 01", end = "2016 12 31",
    freq = "10D")
```

```
DatetimeIndex(['2016-01-01', '2016-01-11', '2016-01-21', '2016-01-31',
               '2016-02-10', '2016-02-20', '2016-03-01', '2016-03-11',
               '2016-03-21', '2016-03-31', '2016-04-10', '2016-04-20',
               '2016-04-30', '2016-05-10', '2016-05-20', '2016-05-30',
               '2016-06-09', '2016-06-19', '2016-06-29', '2016-07-09',
               '2016-07-19', '2016-07-29', '2016-08-08', '2016-08-18',
               '2016-08-28', '2016-09-07', '2016-09-17', '2016-09-27',
               '2016-10-07', '2016-10-17', '2016-10-27', '2016-11-06',
               '2016-11-16', '2016-11-26', '2016-12-06', '2016-12-16',
               '2016-12-26'],
              dtype='datetime64[ns]', freq='10D')

# or 12 hours

>>> pd.date_range(start = "2016  01 01", end = "2016 12 31",
    freq = "12H")

DatetimeIndex(['2016-01-01 00:00:00', '2016-01-01 12:00:00',
               '2016-01-02 00:00:00', '2016-01-02 12:00:00',
               '2016-01-03 00:00:00', '2016-01-03 12:00:00',
               '2016-01-04 00:00:00', '2016-01-04 12:00:00',
               '2016-01-05 00:00:00', '2016-01-05 12:00:00',
               ...

# with frequency on Monday

>>> pd.date_range(start = "2016  01 01", end = "2016 12 31",
    freq = "W-Mon")

DatetimeIndex(['2016-01-04', '2016-01-11', '2016-01-18', '2016-01-25',
               '2016-02-01', '2016-02-08', '2016-02-15', '2016-02-22',
               '2016-02-29', '2016-03-07', '2016-03-14', '2016-03-21',
               '2016-03-28', '2016-04-04', '2016-04-11', '2016-04-18',
               ...
```

```
# with frequency on Wednesday
```

```
>>> pd.date_range(start = "2016  01 01", end = "2016 12 31",
    freq = "W-Wed")
```

```
DatetimeIndex(['2016-01-06', '2016-01-13', '2016-01-20', '2016-01-27',
               '2016-02-03', '2016-02-10', '2016-02-17', '2016-02-24',
               '2016-03-02', '2016-03-09', '2016-03-16', '2016-03-23',
               '2016-03-30', '2016-04-06', '2016-04-13', '2016-04-20',
...
```

There are other methods that allow us to handle dates. Here is another way of creating a range of dates:

```
>>> range1 = pd.date_range(start = "2016  01 01", end = "2016
    03 31", freq = "D")
```

```
# and now we use the .weekday_name method
```

```
>>> range1.weekday_name
```

```
array(['Friday', 'Saturday', 'Sunday', 'Monday', 'Tuesday', 'Wednesday',
       'Thursday', 'Friday', 'Saturday', 'Sunday', 'Monday', 'Tuesday',
       'Wednesday', 'Thursday', 'Friday', 'Saturday', 'Sunday', 'Monday',
       'Tuesday', 'Wednesday', 'Thursday', 'Friday', 'Saturday', 'Sunday',
       'Monday', 'Tuesday', 'Wednesday', 'Thursday', 'Friday', 'Saturday',
...
```

Using Jupyter, we can view the methods for managing dates using the Tab key (Figure 11-1).

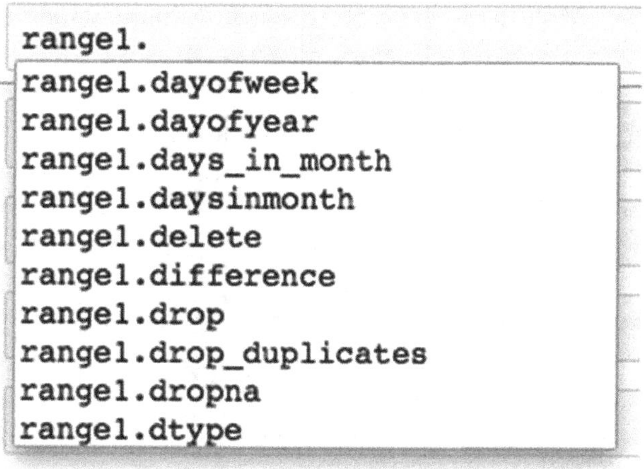

```
range1.
range1.dayofweek
range1.dayofyear
range1.days_in_month
range1.daysinmonth
range1.delete
range1.difference
range1.drop
range1.drop_duplicates
range1.dropna
range1.dtype
```

Figure 11-1. Managing dates with the Tab key

Data Sources

For starting with data mining and machine learning we will use a lot of dataset too understand how an algorithm works. Many data mining datasets can be downloaded from the University of California at Irvine (UCI) online store. The UCI web site (http://archive.ics.uci.edu/ml/index.php) (Figure 11-2) includes all the most used datasets for data science, such as iris, Boston, Wine, SMS Spam collection, and many more (http://archive.ics.uci.edu/ml/datasets.html).

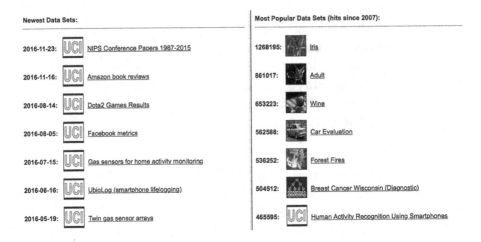

Figure 11-2. *Some of the datasets on the UCI web site*

Recently, even Kaggle has begun to encourage data scientists to publish datasets (https://www.kaggle.com/datasets) to effect exchange among data scientists (Figure 11-3).

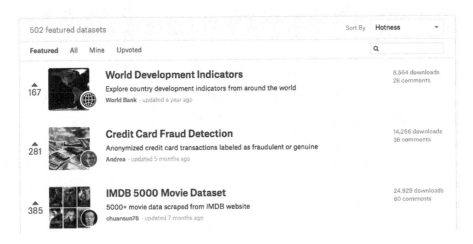

Figure 11-3. *Some of the datasets on the Kaggle webs ite*

As we have seen, the Scikit-learn package also includes datasets that can be imported. For more information on scikit-learn and the featured datasets, you browse the package documentation at `http://scikit-learn.org/stable/datasets/`.

A pandas package module, called *datareader*, features tools to extract data from some online sources (`https://pandas-datareader.readthedocs.io/en/latest/remote_data.html#google-finance`)—particularly those dealing with stock exchange repositories, such as Yahoo! Finance and Google Finance.

Index

© Valentina Porcu 2018
V. Porcu, *Python for Data Mining Quick Syntax Reference*,
https://doi.org/10.1007/978-1-4842-4113-4

E

F

G

H

I

J

K

L

O

P, Q

R

S

Printed in the United States
By Bookmasters